Harry Wirth, Karl-Anders Weiß, Cornelia Wiesmeier
Photovoltaic Modules

Also of Interest

Solar Energy and Technology.
Vol. 1 English-German Dictionary / Deutsch-Englisch Wörterbuch
Mijic, 2016
ISBN 978-3-11-047575-3, e-ISBN 978-3-11-047717-7

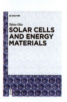
Solar Cells and Energy Materials.
Oku, 2016
ISBN 978-3-11-029848-2, e-ISBN 978-3-11-029850-5

Organic and Hybrid Solar Cells.
An Introduction
Schmidt-Mende, Weickert, 2016
ISBN 978-3-11-028318-1, e-ISBN 978-3-11-028320-4

Physics of Energy Conversion.
Krischer, Schönleber, 2015
ISBN 978-1-5015-0763-2, e-ISBN 978-1-5015-1063-2

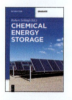
Chemical Energy Storage.
Schlögl (Ed.), 2012
ISBN 978-3-11-026407-4, e-ISBN 978-3-11-026632-0

Energy Harvesting and Systems.
Materials, Mechanisms, Circuits and Storage
Shashank Priya (Editor-in-Chief)
ISSN 2329-8774, e-ISSN 2329-8766

Harry Wirth, Karl-Anders Weiß,
Cornelia Wiesmeier

Photovoltaic Modules

Technology and Reliability

DE GRUYTER

Authors
Dr. Harry Wirth
Fraunhofer ISE
Photovoltaic Modules, Systems and Reliability
Freiburg, Germany

Dr. Karl-Anders Weiß
Fraunhofer ISE
Service Life Analysis
Freiburg, Germany

Dr. Cornelia Wiesmeier
Fraunhofer ISE
Service Life Analysis
Freiburg, Germany

ISBN 978-3-11-034827-9
e-ISBN (PDF) 978-3-11-034828-6
e-ISBN (EPUB) 978-3-11-038398-0
Set-ISBN 978-3-11-034829-3

Library of Congress Cataloging-in-Publication Data
A CIP catalog record for this book has been applied for at the Library of Congress.

Bibliographic information published by the Deutsche Nationalbibliothek
The Deutsche Nationalbibliothek lists this publication in the Deutsche Nationalbibliografie;
detailed bibliographic data are available on the Internet at http://dnb.dnb.de.

© 2016 Walter de Gruyter GmbH, Berlin/Boston
Cover image: Liufuyu/istock/thinkstock
Typesetting: PTP-Berlin, Protago-TeX-Production GmbH, Berlin
Printing and binding: CPI books GmbH, Leck
♾ Printed on acid-free paper
Printed in Germany

www.degruyter.com

Preface

In December 2015, the United Nations Climate Change Conference in Paris sent a clear signal for a worldwide energy transition towards renewables, a transition to take place long before total consumption of fossil and nuclear resources and timely before catastrophic climate change. Photovoltaic (PV) electricity generation is developing into a major global player in our future energy supply scenarios, enabled by a cost reduction momentum not foreseen 15 years ago even by most optimistic PV supporters. Recent bids promise to supply PV electricity at 3 ct($)/kWh in Dubai from a large 800 MW power plant and at 7 ct(€)/kWh in central Europe from small plants below 10 MW. The cost learning curve will keep progressing, since many innovations are well underway in all parts of the value chain. PV modules are the key components for every PV power plant from tiny roof-top systems of a few kW up to plants in the Gigawatt range which may demand millions of modules. Modules are required to efficiently and safely convert solar irradiance into electric power over a service life of decades. The first part of this book addresses crystalline silicon, wafer-based module technology capable to fulfil these expectations. A proper module design and choice of materials requires a thorough understanding of a variety of solar cell properties. With these specifications in mind, module material components and manufacturing issues are elaborated, focusing the key steps of cell interconnection and encapsulation. The cell-to-module loss and gain mechanisms in power and efficiency are discussed in detail, since they are crucial for optimizing the product. After considering nominal module properties displayed under standard testing conditions, module performance in the field under varying ambient conditions is tackled. The second part of the book gives an overview on the very important topic of module reliability which is crucial for the long-term operation and so also for the investments in PV systems. It describes analytical methods for module and material characterization, relevant loads for PV modules and the design of accelerated ageing tests adapted to PV module technology. Also a description is given how reliability tests for materials and modules can be developed. In the end international standards for type approval and safety testing of modules are described as well as their meanings and limitations.

All three authors would like to thank their colleagues at the Fraunhofer Institute for Solar Energy Systems ISE in Freiburg, Germany for many valuable discussions and generous contributions to this book from their various R&D-projects.

Contents

Preface —— V

Symbols and units —— XI

Part I: Crystalline Silicon Module Technology

1 Introduction —— 3

2 Solar cell properties —— 9
2.1 Types of solar cells —— 11
2.2 IV parameters and the electric model —— 11
2.2.1 IV curve —— 13
2.2.2 One-diode model —— 15
2.2.3 Deriving circuit parameters from measured IV curves —— 21
2.2.4 Two-diode model with reverse breakdown —— 23
2.3 Cell efficiency —— 24
2.4 Spectral response —— 25
2.5 Temperature coefficients for cell power —— 28
2.6 Low light response —— 29
2.7 Mechanical properties —— 29
2.8 Thermomechanical properties —— 30
2.9 Cell metallization and contact pads —— 31
2.9.1 Front-to-back contact cells —— 31
2.9.2 Back contact cells —— 35
2.10 Antireflective texturing and coating —— 37

3 Module design, materials, and production —— 41
3.1 Cell interconnection —— 43
3.1.1 Ribbons and wires —— 43
3.1.2 Structured interconnectors —— 47
3.1.3 Conductive backsheet —— 49
3.1.4 Cell shingling —— 50
3.1.5 Soldering processes —— 50
3.1.6 Solders —— 54
3.1.7 Electrically conductive adhesives —— 60
3.1.8 Joint characterization —— 61
3.2 Covers and encapsulants —— 64
3.2.1 Front cover —— 64

3.2.2	Rear cover —— 69	
3.2.3	Encapsulants —— 71	
3.2.4	Edge sealed designs without encapsulant —— 76	
3.2.5	Laminate characterization —— 77	
3.3	Junctions and frame —— 80	
3.4	Module production —— 84	
3.4.1	Production process —— 84	
3.4.2	Production equipment —— 85	
3.5	Module recycling —— 91	
4	**Basic module characterization —— 93**	
4.1	Light IV measurement —— 93	
4.2	Energy rating —— 96	
4.3	Dark IV measurement —— 97	
4.4	Electroluminescence imaging —— 99	
5	**Module power and efficiency —— 103**	
5.1	IV parameters and electric model —— 103	
5.2	Partial shading and hot-spots —— 104	
5.3	Power and efficiency model —— 106	
5.4	Geometrical effects —— 109	
5.5	Optical effects —— 111	
5.5.1	Air/glass and glass/encapsulant interface reflection (f_1, f_3) —— 112	
5.5.2	Glass and encapsulant bulk absorption (f_2, f_4) —— 114	
5.5.3	Active area interface reflection (f_5) —— 115	
5.5.4	Active area multiple reflection (f_6) —— 116	
5.5.5	Finger multiple reflection (f_7) —— 117	
5.5.6	Cell interconnector shading and multiple reflection (f_8) —— 119	
5.5.7	Cell-spacing multiple reflection (f_9) —— 122	
5.6	Electrical effects —— 124	
5.6.1	Cell mismatch (f_{10}) —— 124	
5.6.2	Series resistance losses (f_{11}, f_{12}, f_{13}) —— 125	
5.7	Comprehensive model —— 127	
6	**Module performance —— 131**	
6.1	Irradiation models —— 132	
6.2	Temperature effects —— 136	
6.3	Irradiance level —— 139	
6.4	Irradiance angle —— 139	
6.4.1	Incidence angle modifiers —— 139	
6.4.2	Angular distribution of radiation —— 142	

6.5	Irradiance spectrum —— 143
6.6	Bifacial irradiance —— 144
6.7	Energy payback time —— 146

7 References —— 147

Part II: Crystalline Silicon Module Reliability

8 Characterization of modules and degradation effects —— 155
8.1	Destructive analytics —— 155
8.1.1	Gel content analysis —— 155
8.1.2	Differential scanning calorimetry DSC —— 155
8.1.3	Dynamic mechanical analysis DMA —— 156
8.1.4	Energy-dispersive x-ray spectroscopy EDX —— 156
8.1.5	Auger electron spectroscopy AUGER —— 157
8.1.6	Peel testing —— 158
8.2	Non-destructive analytics —— 158
8.2.1	IV-curve measurements —— 158
8.2.2	Internal and external quantum efficiency IQE and EQE measurements —— 158
8.2.3	Photoluminescence PL imaging —— 159
8.2.4	Gloss- and colour measurement —— 161
8.2.5	Raman spectroscopy —— 162
8.2.6	Nanoindentation —— 167
8.2.7	Fourier-Transform FT-IR/UV/vis measurement —— 167
8.2.8	Scanning acoustic microscopy SAM —— 169

9 Loads for PV Modules —— 171
9.1	External loads —— 171
9.1.1	UV-radiation —— 172
9.1.2	Temperature —— 173
9.1.3	Humidity —— 174
9.1.4	Mechanical loads —— 177
9.1.5	Corrosivity —— 178
9.1.6	Other chemical loads —— 179

10 Accelerated aging tests —— 183
10.1	Light sources —— 184
10.2	Climatic chambers —— 186
10.3	Procedures —— 189

11	**Reliability testing of materials —— 191**
11.1	Equipment —— 191
11.2	Procedures —— 193
12	**Reliability testing of modules —— 195**
12.1	Equipment —— 195
12.2	Procedures —— 195
13	**PV module and component certification —— 197**
13.1	Type approval testing IEC 61215 —— 197
13.2	Safety testing IEC 61730 —— **200**
14	**References —— 203**

Index —— 205

Symbols and units

Symbol	Unit	Description
α	1/m	absorption coefficient
α	°	angle
α_{rel}	1/°C	module short circuit current relative temperature coefficient [78]
A	m²	area, cross section
A	1	absorptance
β	°	angle
β_{rel}	1/°C	module open voltage relative temperature coefficient [78]
δ_{rel}	1/°C	module peak power relative temperature coefficient
d	m	diameter, distance, layer thickness
E	W/m²	irradiance
E	J	electric energy
E_λ	W/m³	spectral irradiance
FF	1	fill factor
H	MJ/m² (kWh/m²)	irradiation (1 kWh = 3.6 MJ)
I	A	current
$K_{\tau\alpha}$	1	incidence angle modifier
L	m	edge length
m_{air}	1	air mass
n	1	refractive index or counting index
n	1	diode ideality factor
P	W	power
PR	1	performance ratio (of PV module or PV power plant)
QE	1	quantum efficiency
R	1	reflectance
r	Ω	sheet resistivity
ρ	Ω·m	electrical (volume) resistivity
SR	A/W	spectral response
T	1	transmittance
T	°C, K	temperature in degree Celsius or in Kelvin as indicated
T_{hom}	1	homologous temperature
V	V	voltage
v	m/s	velocity
w	m	width

Subscripts

AC	alternating current
BB	busbar
beam	beam or direct (irradiance, irradiation)
bulk	referring to the material volume, without interfaces
cell	referring to the cell
d	diode
DC	direct current
e	electron
eff	effective
enc	encapsulant
ext	external
F	finger (part of cell metallization)
glob	global (irradiance, irradiation)
diff	diffuse (irradiance, irradiation)
int	internal
mod	PV module
mpp	maximum power point
OC	open circuit
ph	photon
plant	PV power plant
POA	plane of the array
SC	short circuit or solar constant
STC	standard testing conditions
T	thermal
TIR	total internal reflectance

Abbreviations

AR	anti-reflective
BJBC	back junction back contact
c-Si	crystalline silicon
CTE	coefficient of thermal expansion
CPV	concentrating PV
CTM	cell-to-module
DNI	direct normal irradiation
ECA	electrically conductive adhesives
EL	electroluminescence
EPBT	energy payback time
EVA	ethylene-vinyl acetate
IAM	incidence angle modifier
IBC	interdigitated back contact
ICA	isotropic electrically conductive adhesives
mono-Si	monocrystalline silicon
poly-Si	poly- or multi-crystalline silicon
MWT	metal wrap through
NMOT	nominal module operating temperature
NOCT	nominal operating cell temperature
PDMS	polydimethylsiloxane
PID	potential induced degradation
PV	photovoltaic(s)
PVB	polyvinyl butyral
QA	quality assurance
TCO	total cost of ownership
TMF	thermo-mechanical fatigue
WVTR	water vapor transmission rate

Part I: **Crystalline Silicon Module Technology**

1 Introduction

The first part of this book deals with **wafer-based** crystalline silicon (c-Si) module technology. In terms of market share, c-Si accounts today for about 90 % of the installed PV capacity. The rest is mostly thin film (TF) PV, comprising cadmium telluride (CdTe) and copper indium selenide (CIS) technologies. Concentrating PV (CPV) installations total about 0.2 %. Figure 1.1 also includes installed capacity development of concentrated solar power (CSP).

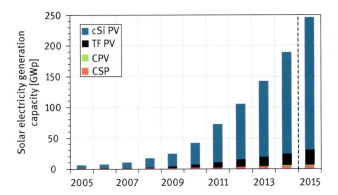

Fig. 1.1: Development of the cumulated solar electricity generation capacity for different PV technologies and for CSP, estimated from different sources, with predicted values for 2015 installations.

Module assembly is the step within the c-Si PV production chain (Figure 1.2) that transforms solar cells into reliable and safe products for electricity generation in PV systems. It also impacts the initial conversion efficiency of the active parts, the solar cells.

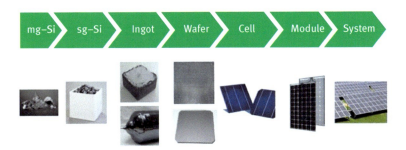

Fig. 1.2: c-Si PV production chain.

Photovoltaic modules directly convert solar into electrical energy at convenient voltage and current levels. The discovery of the photo-electric effect is credited to French scientist Alexandre Edmond Becquerel. He built an electrolytic cell in 1839 that was able to deliver electric power when exposed to light. More than a century later, in 1954, Daryl Chapin, Calvin Fuller, and Gerald Pearson developed a silicon-based PV cell at Bell Laboratories, reporting efficiencies up to 6 % [1]. The left side of Figure 1.3 shows an early PV performance measurement.

Fig. 1.3: Solar battery testing for rural telephone lines by Bell Laboratories engineers in 1955 (left side, reprinted with the permission of Alcatel-Lucent USA Inc) and Vanguard Satellite equipped with 6 modules of about 5 × 5 cm² in size (right side, reprinted with the permission of J. Perlin [2]).

On April 26, one day after their report was published, the New York Times featured the headline "Vast Power of the Sun Is Tapped By Battery Using Sand Ingredient" on its first page and already envisioned solar cells as a potential technology to access the "limitless energy of the sun". A few years later, in 1958, PV space application commenced. Solar cells were interconnected, protected by heavy glass and mounted on a US Vanguard I space satellite to power a radio transmitter (right side of Figure 1.3). The mission proved highly successful, since the PV modules allowed radio operation long after the satellite's conventional batteries faded. Vanguard triggered the success story of PV powered satellites. In the late 1950s, the cost of solar cells was estimated to be in the range of 250–300 US$/W, which is more then 2000 US$/W in today's currency value.

In the 1960s, the potential of PV modules to supply energy in remote applications attracted growing interest, e.g. for lighthouses or offshore navigation signals, where utility grid connection or alternate energy sources were more expensive. Sharp produced modules for lighthouses with cells connected on a circuit board and protected in an acrylic box. These modules achieved 4.5 % nominal efficiency [3].

The oil crisis of 1973 triggered research and development in alternative energy resources worldwide. In the United States, Sandia Laboratories started to develop PV technology for mass production. Several companies like Sharp, Philips and Solar Power introduced modules for terrestrial applications with 5–6 % efficiency. In the first modules from Solar Power, the cells were mounted on rigid substrates and covered by transparent silicone without additional cover.

In 1975, the US government started a series of procurement activities that accelerated technology and testing development for PV modules. First, encapsulated modules with glass covering appeared on the market, followed by the first laminated modules using polyvinyl butyral (PVB) as an encapsulant and a polyester film as a rear cover. The latter design employing the front cover as a rigid support became the predominant module design for the following decades. In the late 1970s, Tedlar was introduced as backsheet material. In the early 1980s, and PVB started to be replaced by ethylene-vinyl acetate (EVA) as encapsulation material.

With decreasing specific cost, in the 1980s PV modules become a convenient power supply for mobile or off-grid consumer devices and small appliances, often combined with an electric battery.

In the 1990s, Germany announced a substantial PV program aiming to implement 1000 roof systems, to be followed by a 100,000 roof program in 1999. With these programs, grid-connected PV started to be recognized as an important candidate for renewable electricity supply. Japan and California also announce PV support schemes in the 1990s.

The German feed-in-tariff (FIT) scheme issued in 2001 boosted PV mass production and initiated tremendous cost reductions (Figure 1.4). After 2010, PV became more and more competitive compared to other electricity generation technologies and reached grid parity in many regions.

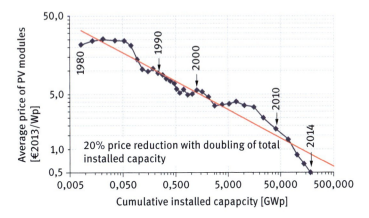

Fig. 1.4: Historical price development of PV modules; the straight line shows the price development trend (PSE AG/Fraunhofer ISE; data from: Strategies Unlimited/Navigant Consulting/EuPD).

Cost reduction was achieved through progress in cell design and in PV production technology, coupled with scaling effects from mass production.

China and Taiwan succeeded in increasing their market share in PV production to about 75 % by the year 2015. This was mainly due to huge investments in large-scale and often integrated production facilities exceeding 1 GW_p annual production capacity (Figure 1.5). In the same period, Europe and the United States lost almost their entire market share, which had reached up to 60 % in the late nineties.

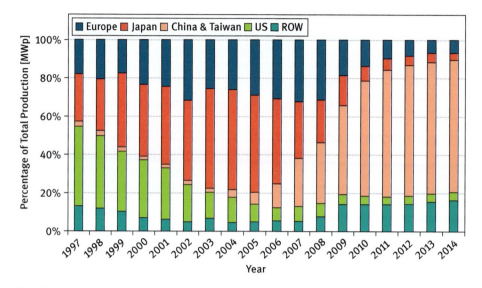

Fig. 1.5: PV production shares by region [4].

Today, modules from mass production using top efficiency cells provide nominal module efficiencies in the range of 19–21 %. These modules can convert one fifth of the solar radiation energy incident on the full module area into electric energy under standard testing conditions. Average commercial module efficiency of newly released modules (Figure 1.6) exceeds 16 % for monocrystalline silicon (mono-Si) cell technology and 15 % for polycrystalline (poly-Si). In field operation throughout the year, common modules perform at about 85–90 % of their nominal efficiency, while bifacial modules may reach performance values above 100 % (Section 6.6).

The dominant market today is **grid-connected** PV. It is supplied with modules typically comprising 60 or 72 solar cells, about 1 m × 1.6 m in size, and connected in series. Their DC voltage is converted to the AC voltage level required by the grid connection point via central, string or module-integrated inverters, possibly complemented by a transformer. Grid-connected PV does not depend on local consumption of electricity.

Off-grid PV may require smaller modules and even cell formats, depending on the application, and is usually part of a system including storage and/or a backup generator, e.g. a diesel generator. These modules may power applications where either the cost of grid connection would be higher, or in remote locations where no grid is available. Developing countries with limited grid ressources generate a huge demand for PV powered solar home systems.

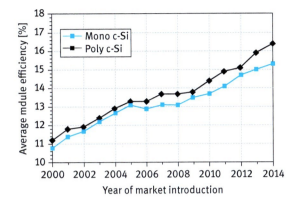

Fig. 1.6: Average module efficiency for newly introduced module types, data from Photon International 2014-02.

The market segmentation for grid-connected PV can be described according to system size and mounting concept. In many countries, the largest share of PV modules is delivered to ground-mounted utility-scale power plants (**utility segment**), typically ranging in size from 5 to 200 MW$_p$. The **industrial segment** comprises roof-top installations often in the range of 1 to 5 MW$_p$. The **commercial segment** includes rooftop or BIPV systems of typically 10–1000 kW$_p$ on warehouses, farm houses, or other commercial property. In the **residential segment** with its rooftop or BIPV installations, system sizes typically are 3–10 kW$_p$. Especially in residential applications, the restricted space availability constitutes an advantage for high efficiency modules.

A PV installation is called **building integrated** PV (BIPV) if the PV modules provide an indispensable functionality to the building, thus replacing conventional building material. If a BIPV module is removed from a building, by definition this would leave the building disfunctional. Most common BIPV applications include back-ventilated curtain wall, semitransparent single or double-glazed façade or roof elements, PV roof tiles, and other specialized modules for roof integration. In BIPV applications, the focus expands from yield and electricity cost to multifunctionality. Besides power generation, the modules may provide shading, partial transparency, or noise protection, and their appearance is subject to esthetic considerations.

Specially designed modules with reduced weight and increased flexibility have also been demonstrated for mobile applications such as ships, trucks, or planes (Figure 1.7). The Turanor Planet Solar launched in 2010 is a catamaran entirely powered by solar energy and is the largest solar vessel worldwide. A 93.5 kW PV system and lithium-ion batteries supply the power for the electric motors. The solar plane Solar Impulse 2 completed in 2014 with a wingspread of 72 m uses a 45 kW PV system. The plane began a circumnavigation of the earth in March 2015. Figure 1.7 shows the Solar Impulse 1 flying over the Golden Gate Bridge. The electric truck E-Force One uses

2000 W of PV modules to supply 5 % of the total energy consumption. The rest of the energy has to be supplied by charging the batteries from the grid.

Fig. 1.7: PV modules fully or partially powering electric vehicles. Reprinted with the permission of Planetsolar, Solar Impulse|Revillard |Rezo, COOP.

In an attempt to reduce space requirements and cost for both PV modules and solar thermal collectors installed side by side, hybrid PV thermal (PVT) modules have been developed. In some configurations, PV laminates are simply glued to a heat exchanger plate with rear side thermal insulation. Water or air is used for heat transfer. Improved thermal efficiencies are reached by adding a glass cover above the PV laminate, possibly in conjunction with leaving off the glass on the PV laminate [5]. In PVT hybrid application, the continuous cell operating temperatures deserves special attention as well as the impact of potentially high stagnation temperatures on the PV components.

2 Solar cell properties

The electric properties of PV modules are governed by their active devices, the **solar cells**. It is therefore crucial to understand single cell properties before looking at their reciprocal interaction in a series circuit and the additional optical and electrical effects inside a module. We will begin with a qualitative look at the cell's functional principle.

Wafer-based solar cells are planar silicon **diodes** that are able to generate voltage and current under illumination. The cell is composed of essentially two layers with different doping of the silicon base material (Figure 2.1). In the very thin top layer of approx. 1–2 µm, the **emitter**, doped with phosphorous donor atoms, provides free electrons as charge carriers. In the bottom layer, the **base**, doping with boron acceptor atoms provides "holes" that can also be regarded as free charge carriers. The outlined doping concept leads to p-type cells; if it is inverted, n-type cells will result.

The left side of Figure 2.1 shows the nonequilibrium state as achieved after doping. As soon as the p and the n layer get in contact, diffusion of positive and negative free charge carriers into the opposite layer sets in immediately. After diffusion, these carriers recombine with some of the majority charge carriers (e.g. holes in the p-type material) and no longer contribute to conductivity. The resulting positively charged donor atoms and negatively charged acceptor atoms build up an electric field across the junction which impedes further diffusion and is associated to a **diffusion voltage** V_D. The electric field creates a depletion region (or space charge region) by quickly removing any free charge carriers from the junction; the associated electric current is called drift current. In the equilibrium state, the potentials that drive diffusion and opposed drift currents cancel out and no current flows. The right side of Figure 2.1 shows this equilibrium state and the corresponding diode symbol.

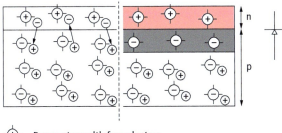

Donor atom with free electron

Acceptor atom with free hole

Fig. 2.1: Schematic drawing of a solar cell showing the opposite doping of the silicon wafer with donors on top and acceptors at the bottom (silicon atoms are not shown). On the left side, their free electrons and holes are depicted in a nonequilibrium condition, before some of them diffuse into the opposite zone. On the right side, the equilibrium condition is shown, after diffusion and recombination of some free charge carriers, and after formation of the depletion region.

Common diodes behave like an insulator for small voltages, since the depletion zone does not offer any substantial free charge carriers. However, if a voltage is applied with the positive polarity to the p region and the negative polarity to the n region, and if this voltage exceeds the **threshold voltage** V_{TH}, approximately equal to the diffusion voltage in the range of 0.6...0.7 V for a silicon diode, the space charge is overcompensated. The space charge region disappears, the diode only shows a very small ohmic resistance, and the current increases substantially. In this case, the diode is operated in **forward bias**.

On the other hand, if the external polarity is reversed, the space charge region increases. The diode is operated in **reverse bias** where it shows very high resistance. Only if the reverse voltage exceeds the diode's **breakdown voltage** V_{BR}, avalanche effects in the electron movement will sharply increase diode conductivity and may eventually destroy it.

Photons with energies above the silicon **bandgap** of 1.12 eV at 300 K, corresponding to a bandgap wavelength of 1100 nm, can be absorbed in a silicon wafer. For irradiance with longer wavelengths, silicon is a transparent material. The penetration depth in silicon depends on the wavelength. Short wavelength photons are absorbed within the first micrometers, while most long wavelength photons (red and infrared) penetrate much deeper.

When a photon is absorbed in silicon, its energy is transferred to an electron. This electron thereby moves from the **valence band** into the **conduction band**. The excited electron and the corresponding hole become mobile and separate. Electrons travel to the n-zone, and holes move into the p-zone. Carriers that make it to the cell contacts without recombination may then sustain an external current which can deliver electric power. In case of a recombination, meaning that the excited electron moves back to the valence band and annihilates a hole, the excitation energy can not be delivered to an external circuit.

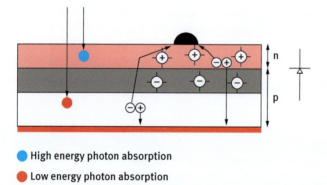

● High energy photon absorption
● Low energy photon absorption

Fig. 2.2: Schematic drawing of a solar cell showing the absorption of different photons (left) and electron-hole-pairs that have been created by such absorption incidents and are separated (right).

Under solar irradiance, electron-hole-pairs are generated throughout the cell area. Separated charge carriers will travel to the cell contacts provided by metallization (Section 2.9). Figure 2.2 shows the black finger cross section on top of the cell and the red full area electrode at the bottom (Section 2.9). Electrons will be conducted laterally to the nearest finger through the n-doped layer (emitter), which provides an increased conductivity. In order to reduce parasitic electron-hole-recombinations at the wafer surface, passivating layers are applied on both sides.

2.1 Types of solar cells

The intense research and development of the last decades generated a variety of wafer-based solar cell technologies. These technologies may differ anywhere along the production chain from the silicon feed stock material purity level over the silicon doping approach, crystallization, wafering and cell format, emitter formation, to passivation and metallization (Table 2.1). While all these aspects are relevant in terms of performance and cost, some of them also require adapted module designs or manufacturing processes.

2.2 IV parameters and the electric model

A solar cell can be described on different levels. The top level addresses the characteristic curve of the device, the so called **IV curve**. This data set of current and voltage value pairs can be measured directly without any knowledge of the internal device structure. The IV curve includes a set of particularly important operation points called **IV parameters**, which refer to maximum cell power and efficiency, short circuit, and open voltage conditions.

For a more profound analysis of the device, a second description level is introduced by using an **equivalent circuit**. This model usually contains a current source and one or more diodes and resistors, which are described by circuit parameters commonly used for electronic components.

A third, even deeper description level uses solid state models and associated material parameters of the device, for instance bandgap energy, carrier concentrations, doping densities, diffusion lengths, drift velocities, etc.

Our focus will be on the first two levels which are particularly important for PV module analysis, namely on the IV parametrization, the circuit parametrization, the relationship between the two, and the influence of externals parameters like irradiance intensity, irradiance spectrum and temperature.

Tab. 2.1: Overview of basic choices for industrial solar cell design and production.

Process step	Basic options
Silicon feed stock production	Process – Siemens reactor – FBR (fluidized bed reactor) – UMG (upgraded metallurgical grade)
Primary doping	p-type, n-type
Crystallization	Process/material – mono-FZ (mono-crystalline float-zone) – mono-Cz (mono-crystalline Czochralski) – mono epitaxial growth (on temporary or permanent substrates) – poly casted (poly-crystalline) – poly EFG (edge-defined film-fed growth) – poly string-ribbon
Wafer formation	Process – sawing – film-fed growth – lift off Format: square, pseudo-square, rectangular
Emitter formation	Process: diffusion, ion implantation, deposition of thin doped layers
Passivation	Design – BSF (back surface field, e.g. aluminum) – PERC (passivated emitter and rear cell) – PERT (passivated emitter and rear totally diffused) – PERL (passivated emitter and rear locally diffused) – HJT (heterojunction technology) Process: diffusion, ion implantation, deposition of thin doped layers Materials: AlSi, SiN_x, SiO_x, aSi, Al_2O_3
Antireflection texturing	Texture – random pyramids (by alkaline etching of mono wafers) – isotropic (by acidic etching of poly wafers) – inverted pyramids (by masked etching) – random needles (by reactive ion etching)
Antireflection coating	Design: single/double coating Materials: SiNx, SiOx
Metallization	Design: front-back-contacts (fbc) or back-contacts (bc) Application: screen-print, galvanic, vapor deposition Materials: silver, copper, aluminum
Active sides	Monofacial, bifacial

2.2.1 IV curve

The **dark IV curve** of a silicon solar cell (Figure 2.3) corresponds to a silicon diode. It shows a forward threshold voltage in the range of +0.7 V, marked by a sharp increase in forward current. The second quadrant in Figure 2.3 displays the reverse bias operation domain of the solar cell. When the voltage approaches the reverse breakdown voltage V_{BR} of −14 V in this case, the current steeply increases. Diode characteristics of the solar cell may vary over the area, which leads to a spacially inhomogeneous V_{BR} and associated breakdown behavior. In areas with lowest V_{BR}, high current densities may arise accompanied by high temperatures, so-called **hot-spots**. This behavior is critical in PV modules where many serially interconnected cells can generate substantial reverse voltages (Section 5.2).

When the solar cell is illuminated (E > 0), the light induced current shifts the IV-curve upwards. The cell delivers power when the product $P = I * V$ is positive, which corresponds to the red-rimmed first quadrant. At negative power values, in the second quadrant, the cell dissipates power.

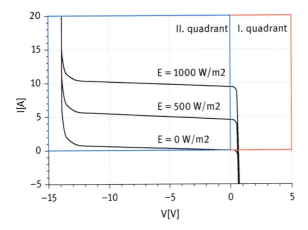

Fig. 2.3: Calculated IV-curves of a solar cell for different irradiance levels E including the dark IV curve (E = 0 W/m²).

Solar cells are intended to operate within the first quadrant, where their IV parameters are defined (Figure 2.4). The **open circuit voltage** (V_{OC}) and **short circuit current** (I_{SC}) are easily readable at zero current and zero voltage, respectively. Theoretically achievable V_{OC} for a silicon solar cell at 1 sun is 0.85 V, whereas lab cells today reach maximum values of 0.75 V [6]. The short circuit current depends on the short circuit current density J_{SC} and the cell area. Best laboratory J_{SC} values exceed 40 mA/cm².

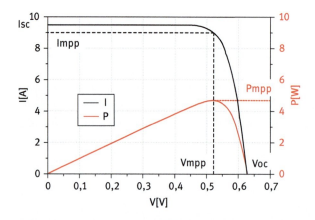

$I_{sc}[A]$	$V_{oc}[V]$	$I_{mpp}[A]$	$V_{mpp}[V]$	FF	$P_{mpp}[W]$	η
9.50	0.627	9.00	0.522	79.0%	4.70	19.3%

Fig. 2.4: Calculated IV-curve and power curve of a solar cell and corresponding IV parameters, power and efficiency.

Cell power is given by the product of current I and voltage V, displayed as red curve. The peak of the red curve is called the **maximum power point** (mpp) and indicates maximum power, P_{mpp}, the respective current, I_{mpp}, and voltage, V_{mpp}.

The **fill factor**, as defined by equation (2.1), is useful for the efficiency analysis of solar cells. Industrial solar cells achieve fill factor values in the range of 76–79 %, while high efficiency lab cells achieve 81–83 %

$$FF = \frac{V_{mpp} \cdot I_{mpp}}{V_{OC} \cdot I_{SC}} = \frac{P_{mpp}}{V_{OC} \cdot I_{SC}}, \qquad (2.1)$$

where
FF   fill factor,
V_{mpp} maximum power point voltage [V],
I_{mpp} maximum power point current [A],
V_{OC} open circuit voltage [V],
I_{SC} short circuit current [A],
P_{mpp} maximum power [W].

The **efficiency** η of a solar cell is defined as the ratio of the delivered electric power [W] at mpp to the radiant flux on the full cell area under a defined irradiance (equation (2.2)). More specifically, the nominal cell efficiency refers to the standard test conditions (STC). These conditions imply perpendicular irradiance of 1000 W/m^2, a

2.2 IV parameters and the electric model

cell temperature of 25 °C and an AM 1.5 spectrum as defined in IEC 60904-3 [7]:

$$\eta = \frac{P_{mpp}}{\Phi_e} = \frac{P_{mpp}}{E \cdot A_{cell}},\qquad(2.2)$$

where
Φ_e radiant flux [W],
P_{mpp} maximum power [W],
A_{cell} cell area [m²],
E irradiance [W/m²].

2.2.2 One-diode model

Several models have been proposed to reproduce basic electric characteristics of solar cells in simple equivalent electric circuits. One difficulty of modelling a real cell arises from the fact that over the entire cell area the characteristics may vary. These variations originate from the silicon crystallization process and from different steps of the cell manufacturing process. The resulting heterogeneous device shows properties that are difficult to capture precisely in simple 1-dimensional models.

In the single diode model (Figure 2.5), I_L denotes the **light induced current**, I_D is the **diode current**, and two resistors account for parasitic effects. R_S is introduced to model **series resistance** losses due to contact and volume resistance. R_P (or R_{SH}), the **parallel resistance** (or **shunt resistance**), accounts for shunting losses within the pn-junction area or at the cell edge and usually originates from cell manufacturing defects.

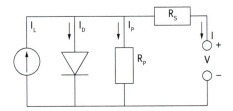

Fig. 2.5: Equivalent circuit according to the single-diode model including series and parallel resistance.

The light-induced current is assumed to be proportional to the irradiance. The diode current I_D depends on the voltage across the diode, V_D, according to the **Shockley diode equation** (2.3):

$$I_D = I_0 \left(\exp\left(\frac{V_D}{nV_T}\right) - 1 \right), \quad V_T = \frac{kT}{q},\qquad(2.3)$$

where

I_D	diode current [A],
I_0	diode saturation current [A],
V_D	diode voltage [V],
n	diode ideality factor,
V_T	thermal voltage [V],
k	Boltzmann constant (8.617×10^{-5} eV/K),
T	temperature [K],
q	elementary charge (1 e = 1.602×10^{-19} C).

I_0, the (reverse) saturation current of the diode, originates from minority carriers (e.g. electrons in the p-region) that recombine in the depletion region. I_0 limits the current in the reverse bias operation mode. The **ideality factor** n can take a value of 1 or 2, depending on the dominant recombination mechanism. V_T denotes the thermal voltage with a value of approx. 26 mV at room temperature. The temperature T has to be inserted as a Kelvin value, which lies 273.15 above the value in °C. The STC temperature of 25 °C thus corresponds to nearly 300 K.

Typical industrial solar cells display an area-normalized series resistance of the order of 1 Ωcm² and an area-normalized parallel resistance of the order of 10 kΩcm². Area-normalized resistance refers to a current density instead of a current and is defined as the product of resistance [Ω] and area [cm²]. For a full square cell in the 156-mm-format, the mentioned area-normalized values correspond to absolute values of 4 mΩ for R_S and 40 Ω for R_P. The R_S and R_P resistors cannot be localized within the solar cell. They should be looked at as lump model parameters that summarize a multitude of different effects within a certain operation range of the cell. In consequence, different operating points of the cell, e.g. due to varying irradiance levels or temperatures, will lead to somewhat different resistance values that best reproduce the measured IV curve of the cell.

The current I and voltage V delivered to an external load according to the one-diode-model (Figure 2.5) is given by the implicit equation (2.4). Its structure follows from application of Kirchhoff's current law to Figure 2.5. The diode voltage V_D has been expressed as the sum of the voltage drop over R_S and the external voltage V:

$$I = I_L - I_D - I_P = I_L - I_0 \left(\exp\left(\frac{V + IR_S}{nV_T} \right) - 1 \right) - \frac{V + IR_S}{R_P}; \quad V_T = \frac{kT}{q}. \quad (2.4)$$

In order to display an IV curve as defined by the implicit equation (2.4) for a given set of parameters (I_L, I_0, n, R_S, R_P) or to compare it with a measured curve, value pairs [V, I(V)] are required. These value pairs for voltages between 0 V and V_{OC} can be found by Newton's method. To apply this method, we define a function $f_V(I)$ with the voltage V as constant parameter and variable current I (equation (2.5)). For each voltage point V, the current I that solves equation (2.4) will be one root of $f_V(I)$:

$$f_V(I) = I - I_L + I_0 \left(\exp\left(\frac{V + IR_S}{nV_T} \right) - 1 \right) + \frac{V + IR_S}{R_P}. \quad (2.5)$$

Roots of $f_V(I)$ are approached by iteration. A suitable starting point for each iteration is the root that was found at the previous voltage value. For the very first iteration at $V = 0$, the light induced current I_L can be used as starting point. Following Newton's method, the root is approached stepwise according to equation (2.6):

$$I_{i+1} = I_i - \frac{f_V(I_i)}{f'_V(I_i)}. \tag{2.6}$$

The derivation of $f_V(I)$ is given by equation (2.7):

$$f'_V(I) = 1 + I_0 \exp\left(\frac{V + IR_S}{nV_T}\right)\left(\frac{R_S}{nV_T}\right) + \frac{R_S}{R_P}. \tag{2.7}$$

The iteration usually converges sufficiently after a few steps. It has to be performed for every voltage point between 0 V and V_{OC}. An estimation for V_{OC} will be derived below; the result from equation (2.26) is rearranged to equation (2.8):

$$V_{OC} \approx nV_T \ln\left(\frac{I_L}{I_0}\right). \tag{2.8}$$

Following this procedure, we study the influence of external parameters and circuit parameters on the shape of the IV curve and the IV parameters according to the model in equation (2.4).

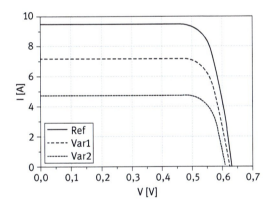

	External par.		Circuit parameters				IV parameters						
Var.	E	Temp	I_L	I_0	R_S	R_P	I_{SC}	V_{OC}	I_{mpp}	V_{mpp}	FF	P_{mpp}	η
Unit	[W/m²]	[°C]	[A]	[nA]	[mΩ]	[Ω]	[A]	[V]	[A]	[V]	[%]	[W]	[%]
Ref.	1000	25	9.50	0.24	3	20	9.50	0.627	8.98	0.524	79.0	4.71	19.3
1	750	25	7.13	0.24	3	20	7.12	0.619	6.78	0.520	79.8	3.52	19.3
2	500	25	4.75	0.24	3	20	4.75	0.609	4.48	0.520	80.4	2.33	19.1

Fig. 2.6: Effect of irradiance change on cell parameters; variations with respect to the reference case are marked red.

If the irradiance is reduced (Figure 2.6), the light induced current decreases proportionally. According to equation (2.8), V_{OC} decreases logarithmically, from which an efficiency loss with decreasing irradiance can be expected. Since the ratio I_L/I_0 is of the order of 10^{10}, an irradiance decrease from 1000 W/m² to 10% or to 1% of this value only affects V_{OC} by a reduction of the order of 10% or 20% relative. For cells with lower parallel resistance, a strong decline of efficiency with irradiance is observed, which may even approach a linear decrease for very low R_P [8]. Considering the equivalent circuit, the power loss in R_P, which equals V_D^2/R_P, only changes very slowly with decreasing irradiance. With strongly decreasing irradiance and cell power, the power loss in R_P thus begins to dominate cell behavior and strongly affects cell efficiency.

On the other hand, high series resistances combined with a moderate parallel resistance may lead to an intermediate increase of FF and efficiency when the irradiance is moderately reduced below 1000 W/m². In this case, the series resistance power loss proportionally decreasing with the square of the current initially dominates efficiency changes.

Increasing temperature (Figure 2.7) will substantially increase the diode saturation current I_0 (equation (2.9); [9]) and only slightly increase the light induced current I_L (equation (2.10); [10]):

$$I_0 \propto T^3 \cdot \exp\left(-\frac{E_G(T)}{kT}\right), \tag{2.9}$$

where

T temperature [K],
E_G energy difference (gap) between the valence band and the conduction band.

As a result, rising temperature negatively affects the open-circuit voltage V_{OC} (equation (2.10)) and thereby the cell power and efficiency:

$$I_{SC} \approx I_L \propto E \cdot (1.0006) \wedge \left(\frac{T - T_{room}}{1\,°C}\right),$$
$$V_{OC} \approx V_{OC,T_{room}} + c_1 \cdot (T - T_{room}), \tag{2.10}$$

where

T temperature [°C],
T_{room} room temperature (25 °C),
E irradiance [W/m²],
c_1 temperature coefficient of V_{OC} (−2.3 mV/°C).

The temperature dependency of the fill factor can be calculated using equation (2.11) [9]. With equation (2.1), the cell power can be expressed by the fill factor, and thereby the temperature coefficient of cell power can be expressed:

$$FF = \frac{\frac{V_{OC}}{V_T} - \ln\left(\frac{V_{OC}}{V_T} + 0.72\right)}{\frac{V_{OC}}{V_T} + 1}, \tag{2.11}$$

$$P_{mpp} = V_{OC} \cdot I_{SC} \cdot FF. \tag{2.12}$$

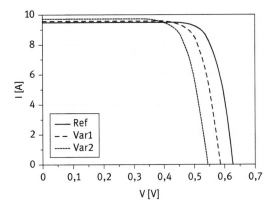

	External par.		Circuit parameters				IV parameters						
Var.	E	Temp	I_L	I_0	R_s	R_p	I_{SC}	V_{OC}	I_{mpp}	V_{mpp}	FF	P_{mpp}	η
Unit	[W/m²]	[°C]	[A]	[nA]	[mΩ]	[Ω]	[A]	[V]	[A]	[V]	[%]	[W]	[%]
Ref.	1000	25	9.50	0.24	3	20	9.50	0.627	8.98	0.524	79.0	4.71	19.3
1	1000	45	9.61	5.11	3	20	9.61	0.585	9.06	0.479	77.1	4.34	17.8
2	1000	65	9.73	76.50	3	20	9.73	0.544	9.10	0.436	74.9	3.96	16.3

Fig. 2.7: Effect of temperature change on cell parameters; variations with respect to the reference case are marked red (temperature coefficients of R_s and R_p are neglected).

From these equations it follows that high V_{OC} values reduce the power temperature coefficient and thus reduce efficiency losses at elevated operation temperatures. Reported power temperature coefficients for different industrial cell technologies are listed in Section 2.5.

Increasing series resistance shows little effect on the short circuit current, but severe effects on fill factor, power, and efficiency (Figure 2.8).

In order to estimate this effect, we define $P_{mpp,0}$ as the mpp cell power for zero series resistance and estimate power loss from series resistance by $R_S I_{mpp,0}^2$ as shown in equation (2.13):

$$P_{mpp} = P_{mpp,0} - R_S I_{mpp,0}^2 = P_{mpp,0}\left(1 - R_S \frac{I_{mpp,0}}{V_{mpp,0}}\right) \approx P_{mpp,0}\left(1 - R_S \frac{I_{SC}}{V_{OC}}\right). \quad (2.13)$$

From equation (2.13), it follows analogously for the influence of R_S on the idealized fill factor FF_0 (with zero series resistance) that

$$FF = \frac{P_{mpp}}{V_{OC} I_{SC}} = FF_0\left(1 - R_S \frac{I_{mpp,0}}{V_{mpp,0}}\right) \approx FF_0\left(1 - R_S \frac{I_{SC}}{V_{OC}}\right). \quad (2.14)$$

Reducing the parallel (or shunt) resistance will also mainly affect fill factor, power, and efficiency. Analogue to this, equations (2.15) and (2.16) can be derived to estimate

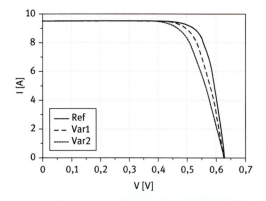

	External par.		Circuit parameters				IV parameters						
Var.	E	Temp	I_L	I_0	R_S	R_P	I_{SC}	V_{OC}	I_{mpp}	V_{mpp}	FF	P_{mpp}	η
Unit	[W/m²]	[°C]	[A]	[nA]	[mΩ]	[Ω]	[A]	[V]	[A]	[V]	[%]	[W]	[%]
Ref.	1000	25	9.50	0.24	3	20	9.50	0.627	8.98	0.524	79.0	4.71	19.3
1	1000	25	9.50	0.24	6	20	9.50	0.627	9.00	0.496	75.0	4.46	18.3
2	1000	25	9.50	0.24	9	20	9.50	0.627	8.94	0.473	71.0	4.22	17.4

Fig. 2.8: Effect of R_S change on cell parameters; variations with respect to the reference case are marked in red.

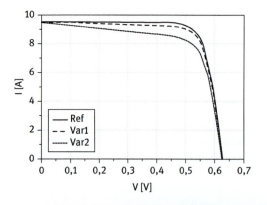

	External par.		Circuit parameters				IV parameters						
Var.	E	Temp	I_L	I_0	R_S	R_P	I_{SC}	V_{OC}	I_{mpp}	V_{mpp}	FF	P_{mpp}	η
Unit	[W/m²]	[°C]	[A]	[nA]	[mΩ]	[Ω]	[A]	[V]	[A]	[V]	[%]	[W]	[%]
Ref.	1000	25	9.50	0.24	3	20.0	9.50	0.627	8.98	0.524	79.0	4.71	19.3
1	1000	25	9.50	0.24	3	2.0	9.49	0.626	8.75	0.524	77.2	4.58	18.8
2	1000	25	9.50	0.24	3	0.5	9.44	0.623	8.04	0.520	71.0	4.18	17.2

Fig. 2.9: Effect of R_P change on cell parameters; variations with respect to the reference case are marked in red.

the effect:

$$P_{mpp} = P_{mpp,0} - \frac{V_{mpp,0}^2}{R_P} = P_{mpp,0}\left(1 - \frac{1}{R_P}\frac{V_{mpp,0}}{I_{mpp,0}}\right) \approx P_{mpp,0}\left(1 - \frac{1}{R_P}\frac{V_{OC}}{I_{SC}}\right), \quad (2.15)$$

$$FF = \frac{P_{mpp}}{I_{SC}V_{OC}} = FF_0\left(1 - \frac{1}{R_P}\frac{V_{OC}}{I_{SC}}\right). \quad (2.16)$$

A reduced shunt resistance will reduce the slope of the IV-curve and negatively affect FF and Pmpp (Fig. 2.9).

2.2.3 Deriving circuit parameters from measured IV curves

The implicit equation (2.4) contains five unknown parameters (I_L, I_0, n, R_S, R_P) that need to be determined. If five different (I, V) value pairs from the cell's measured IV curve are available, five nonlinear equations can be formulated and solved numerically. Two equations can be gained from the end points of the IV curve, namely (I_{SC}, 0) and (0, V_{OC}), and three others from the maximum power point (I_{mpp}, V_{mpp}) and its vicinity.

The procedure can achieve a very close fit around the selected IV points, but not necessarily to the entire IV curve. Better parameter values can be determined by using the least square method or the Lagrange multiplier method. Following the least square method, the parameter set is varied to minimize the sum of squared deviations between calculated and measured values over the entire IV curve or selected sections.

Numerical methods require an initial set of values which is progressively improved by iteration. To obtain reasonable initial values for (I_L, I_0, n, R_S, R_P), approximate analytic solutions of equation (2.4) within limited domains prove helpful.

The short circuit current I_{SC} can be used as approximation and starting point for I_L. The values are very similar for reasonably low series resistance R_S.

For large currents close to I_{SC}, the diode current I_D in equation (2.4) becomes negligible, leaving a simplified equation

$$I = I_L - I_P = I_L - \frac{V + IR_S}{R_P}. \quad (2.17)$$

By differentiating both sides with respect to the voltage V, we obtain

$$\frac{dI}{dV} = -\frac{1}{R_P} - \frac{R_S}{R_P}\frac{dI}{dV}. \quad (2.18)$$

Solving for dI/dV and expressing the reciprocal derivation, dV/dI yields

$$\frac{dV}{dI} = -(R_P + R_S). \quad (2.19)$$

Since the parallel resistance is much higher than the series resistance, it can be approximated by the negative slope of the voltage with respect to the current at I = I_{SC}:

$$R_P = -\frac{dV}{dI}\bigg|_{V=0}. \quad (2.20)$$

In the opposite limit, for large voltages appoaching V_{OC}, the external current becomes very small. The diode is operated close to its threshold voltage V_{TH} and will absorb a much higher current than the parallel resistor. If I_P is neglected, equation (2.4) can be simplified to

$$I = I_L - I_D = I_L - I_0 \left(\exp\left(\frac{V + IR_S}{nV_T} \right) - 1 \right). \tag{2.21}$$

Differentiation with respect to I leads to equation (2.22):

$$1 = -I_0 \exp\left(\frac{V + IR_S}{nV_T} \right) \cdot \frac{\frac{dV}{dI} + R_S}{nV_T}. \tag{2.22}$$

By solving for dV/dI and inserting I = 0 and V = V_{OC} we get

$$\frac{dV}{dI} = -\frac{nV_T}{I_0} \exp\left(-\frac{V_{OC}}{nV_T} \right) - R_S. \tag{2.23}$$

Since V_{OC} is more than 20 times larger than V_T, the first term can be neglected, leading to

$$R_S = -\left.\frac{dV}{dI}\right|_{I=0}. \tag{2.24}$$

Figure 2.10 illustrates the approximate determination of R_S and R_P from the cell's measured IV curve. Although the approximated values for R_P from equation (2.20) and R_S from equation (2.24) may considerably differ from the final solution, they provide convenient starting points for numerical procedures.

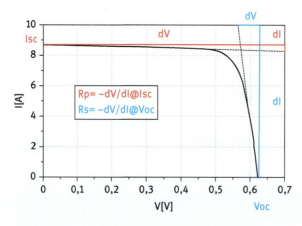

Fig. 2.10: IV curve with tangents (dotted lines) in the I_{SC} and V_{OC} end points; the inverse slope of these tangents provide approximate values for R_P and R_S.; R_S is chosen unrealistically high in this figure for demonstration purposes.

In order to find an initial value for the diode saturation current I_0, equation (2.4) is evaluated at V_{OC} with $I = 0$:

$$0 = I_L - I_0 \left(\exp\left(\frac{V_{OC}}{nV_T}\right) - 1 \right) - \frac{V_{OC}}{R_P}. \tag{2.25}$$

Solving for I_0 and observing that the exponential term is much larger than 1 leads to

$$I_0 = \left(I_L - \frac{V_{OC}}{R_P}\right) \cdot \exp\left(-\frac{V_{OC}}{nV_T}\right) \approx I_L \exp\left(-\frac{V_{OC}}{nV_T}\right). \tag{2.26}$$

If we assume $n = 1$ as a starting value and $I_L = I_{SC}$, for a typical V_{OC} of 0.6 V the initial guess for the diode current will be 10–11 orders of magnitude smaller than the short circuit current. With this, the set of initial values for starting a numerical fitting procedure is complete.

2.2.4 Two-diode model with reverse breakdown

A more precise modeling is possible by introducing a second diode. The first diode maps bulk and surface recombinations with an ideality factor of 1. The second diode maps the recombination current in the junction with $n = 2$, originating from intermediate level Shockley–Read–Hall recombinations. In order to reproduce the breakdown behavior, the parallel resistor needs to be replaced by a variable resistor with a nonohmic characteristic [11, 12]. The resulting 2-diode model is represented by equation (2.27), and the equivalent circuit is shown in Figure 2.11:

$$\begin{aligned}I &= I_L - I_{D1} - I_{D2} - I_P = \\ &= I_L - I_{01}\left(\exp\left(\frac{V_{int}}{V_T}\right) - 1\right) - I_{02}\left(\exp\left(\frac{V_{int}}{2V_T}\right) - 1\right) - \frac{V_{int}}{R_{P,BR}}\end{aligned} \tag{2.27}$$

where

$$V_T = \frac{kT}{q}; \quad V_{int} = V + IR_S; \quad R_{P,BR} = R_P\left(1 + a\left(1 - \frac{V_{int}}{V_{BR}}\right)^{-b}\right)^{-1}.$$

The diode behavior is determined by the saturation current I_{01} and I_{02}. V_{int} is introduced to simplify the equation and corresponds to the voltage across the current source, respectively the junction. The factor a scales the breakdown current linearly. The breakdown exponent b controls the nonlinear behavior of the current in the vicinity of the breakdown voltage V_{BR}. The curves in Figure 2.3 were generated from equation (2.27) with the parameters given in Table 2.2. Detailed derivation and discussion of model parameters can be found in [10, 13].

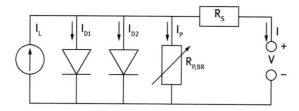

Fig. 2.11: Equivalent circuit according to the two-diode-model, including series and variable parallel resistor.

Tab. 2.2: Parameters used for two-diode model calculations with reverse breakdown.

Value	Unit	Parameter
9.5	A	Light-induced current I_L at 1000 W/m² and 25 °C
0.24	nA	First diode saturation current I_{01} at 25 °C
240	nA	Second diode saturation current I_{02} at 25 °C
0.003	Ohm	Series resistance R_S
20	Ohm	Parallel (shunt) resistance R_P
−14	V	Breakdown voltage V_{BR}
0.05	1	Breakdown scaling factor a
1.1	1	Breakdown exponent b
25	°C	Temperature T

2.3 Cell efficiency

Single-junction silicon solar cells have a theoretical efficiency limit of 29.4 % [14], the so-called Shockley–Queisser limit [15]. For **double-junction** silicon solar cells, the theoretical limit climbs to 42.5 % [16]. In 2014, the efficiency record for full-size single-junction laboratory cells reached 25.6 ± 0.5 % for a **hetero-junction** technology [6]. Top manufacturers have reached 24 % efficiency in mass production, while many other manufacturers achieved 19–20 %.

Solar cells with a full area aluminum back surface field passivation, which has been dominating the market for a several years, cannot make use of light incident from the rear side, they are "monofacial". If dielectric instead of metallic passivation is used on the cell rear side instead, cells may be designed to be responsive from both sides ("bifacial"). Bifacial cells often display a somewhat lower efficiency for rear side illumination. Their efficiency under two-sided illumination depends on the chosen irradiance levels on each side.

Some solar cells lose efficiency after first exposure to light. This light induced degradation (LID) is ascribed to boron-oxygen (B-O) defects within the silicon wafer. Stabilization requires 10–20 MJ/m² (a few kWh) of irradiation and may reduce initial cell nominal power by up to a few percent relative.

2.4 Spectral response

The **external quantum efficiency** QE of a solar cell is defined as the ratio of the number of excited electrons n_e reaching the cell contacts to the number of photons n_{ph} incident on the cell area within a finite wavelength interval $\Delta\lambda$ centered around the wavelength λ (equation (2.28)):

$$QE(\lambda) = \frac{n_e}{n_{PH}(\lambda, \Delta\lambda)}. \tag{2.28}$$

An ideal solar cell would display a QE very close to 100 % over a wavelength range that is limited on the infrared side by the band gap wavelength λ_g of 1100 nm (equation (2.29)) which corresponds to the material's bandgap E_g (1.12 eV at 300 K):

$$\lambda_g = \frac{hc}{E_g}, \tag{2.29}$$

where
h = Planck's constant (4.136×10^{-15} eV · s),
c = speed of light in vacuum (2.998×10^8 m/s).

Real cells lose efficiency due to recombinations at the cell surface and in the bulk material, due to front side reflection losses, transmission losses of the wafer, and absorption losses caused by the cell metallization. If only those photons are considered that are absorbed in the silicon wafer, the ratio of the collected electrons to absorbed photons yields the **internal quantum efficiency** QE_{int}. It can be derived from QE by corrections for reflectance, transmittance, and parasitic absorptance. The latter is mainly caused by front- and rear-side metallization. Solar cells with full area aluminum screen printed rear side will not transmit radiation due to their opaque rear side metallization:

$$QE_{int}(\lambda) = \frac{QE(\lambda)}{1 - R - T - A_{par}}, \tag{2.30}$$

where
QE_{int} internal quantum efficiency [6],
R cell reflectance,
T cell transmittance,
A_{par} cell parasitic absorptance.

The **external spectral response** SR of a solar cell at a wavelength λ is defined as the ratio of the spectral short circuit current $I_{SC,\lambda}$ to the incident spectral irradiance E_λ (equation (2.31)) and the cell area.

$$SR(\lambda) = \frac{I_{sc,\lambda}(\lambda)}{A \cdot E_\lambda(\lambda)} = \frac{q \cdot n_e}{h\nu \cdot n_{PH}} = \frac{q\lambda}{hc} \cdot QE(\lambda), \tag{2.31}$$

where

SR   external spectral response [A/W],
λ   wavelength [m],
$I_{SC,\lambda}$ spectral short circuit current [A/m],
A   cell area [m^2],
E_λ spectral irradiance [W/m^3],
q   elementary charge (1.602×10^{-19} C).

The **spectral irradiance** in wavelength, E_λ, is defined as the derivative of the irradiance with respect to the wavelength (equation (2.32)). The spectral short circuit current $I_{SC,\lambda}$ is defined similarly as the derivative of the short circuit current with respect to the wavelength:

$$E_\lambda = \frac{\partial E}{\partial \lambda}; \quad I_{SC,\lambda} = \frac{\partial I_{SC}}{\partial \lambda}. \quad (2.32)$$

The SR can be measured directly by subjecting the cell to monochromatic irradiance of different wavelengths. An ideal cell with QE very close to 100 % would display a spectral response that linearly decreases with the wavelength, according to equation (2.31). The reason behind this is that even though photons with shorter wavelengths carry more energy than bandgap photons, they can only generate one free charge pair. Their surplus energy is converted to heat.

The short circuit current of a solar cell can be calculated by integrating the external spectral response and the spectral irradiance over the relevant spectral range (equation (2.33)):

$$I_{sc} = A \cdot \int_{300nm}^{1200nm} SR(\lambda) \cdot E_\lambda \, d\lambda. \quad (2.33)$$

For convenient shape comparison, the SR is often normalized with respect to its maximum value over the spectrum, denoted as SR_{norm}. Figure 2.12 shows quantum efficiencies and normalized spectral response curves for an ideal solar cell and measured values from a commercial cell. Due to the semiconductor properties of silicon (the transition of electrons from the valence band into the conduction band involves phonons), QE shows a steep decrease around λ_g but does not fall to zero at λ_g.

The spectral response of the solar cells in the module has implications on the requirements for cover and encapsulation materials. For maximum module performance, the material stack in front of the solar cell must exhibit maximum transmittance for wavelengths where the solar cell is responsive and solar irradiance is available. In practice, the material stack applied in module production will usually change the initial spectral response of the cell due to interface modifications and additional bulk material absorption.

The spectrum irradiated by the sun in the UV, visible and infrared region is very similar to the **black body** radiation of a temperature of 5800 K as shown in Figure 2.13.

At an average distance from the sun, the earth receives above the atmosphere an irradiance of approximately 1366 W/m^2, the **solar constant**. On its way through the

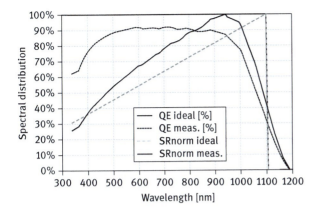

Fig. 2.12: Quantum efficiency QE and normalized spectral response SR for an ideal and a real solar cell. As a result of normalization, the SR curve integrals are not comparable.

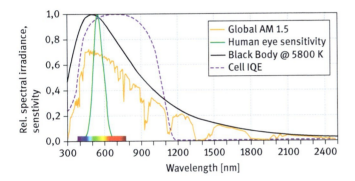

Fig. 2.13: Relative spectral irradiance of a black body at 5800 K and of the AM 1.5 solar spectrum, human eye sensitivity, and internal quantum efficiency of a silicon solar cell.

atmosphere, the solar radiation spectrum changes due to absorption from ozon, water, carbon dioxide, and dust particles, also due to wavelength-dependent scattering. The length of this way, which depends on the solar elevation angle and the observer's location height, also influences the received irradiance. To overcome these variations and uncertainties in the characterization of solar cells and modules, standard spectra have been defined. Figure 2.13 also displays the global solar spectrum AM 1.5 (or AM 1.5g; "g" stands for global) as defined in the standard IEC 60904-3 Edition 2 [7]. This spectrum corresponds to a solar elevation angle of 42° and is predominantly used for laboratory module characterization. In outdoor module operation, the irradiance intensity and spectrum changes with the sun's elevation and with sky condition.

Figure 2.14 shows SR_{norm} and the product of both spectra, $AM1.5 \times SR_{norm}$. This reference spectrum will be further used for the assessment of effective spectral prop-

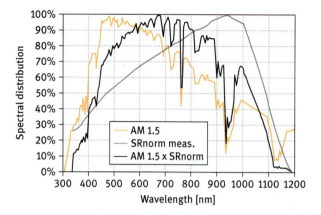

Fig. 2.14: AM 1.5 solar spectrum, normalized measured internal spectral response SR$_{norm}$ for a cell and the product of AM 1.5 and SR$_{norm}$.

erties with respect to the solar spectrum and to an exemplary solar cell. For any specific analysis, the SR of the cell in question should be used instead.

2.5 Temperature coefficients for cell power

PV modules operate within a wide range of temperatures, depending particularly on current solar irradiance, air temperature, wind speed, and mounting. Module power decreases with temperature, due to a dominating negative temperature coefficient of cell voltage. Ohmic resistivity also increases with temperature, leading to increased series resistance losses in metals. The resulting temperature coefficient for module power typically lies between −0.3 to −0.5 %/°C for c-Si technologies (Figure 2.15).

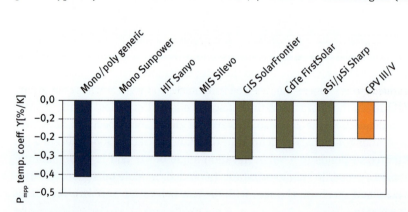

Fig. 2.15: Temperature coefficients of P$_{mpp}$ power for PV modules of different technologies. Source: manufacturer's data sheets, typical values.

High-efficiency solar cells with large V_{oc} tend to lose less power at elevated temperatures.

Nominal module power and efficiency are determined at 25 °C, according to IEC 61215. In warm climates, elevated operation temperatures noticeably reduce the effective module efficiency.

2.6 Low light response

Nominal cell power is reported at an irradiance of 1000 W/m² following standard testing conditions. At lower irradiance levels, the short circuit current decreases linearly. The open circuit voltage decreases logarithmically with irradiance (Section 2.2.2). Additionally, cells with low parallel (shunt) resistance tend to show poor response under low light condition [17, 18]. This behavior can be understood from the diode model (Section 2.2.2). As a consequence of voltage loss, R_P and recombination current dominance, the cell efficiency is reduced. Figure 2.16 shows exemplary low light response data sets from module manufacturer's datasheets. In many locations, frequent cloudiness leads to substantial operation periods of PV modules at irradiance levels far below 1000 W/m². Thus, the low light response can have a strong influence on the module performance (Chapter 6).

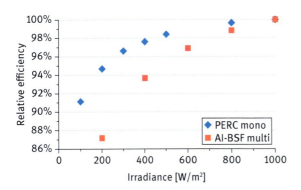

Fig. 2.16: Measured low light response for two commercial modules with different cell technologies (Fraunhofer ISE).

2.7 Mechanical properties

Solar cells are mostly processed on square poly- or mono-Si wafers with 156 mm edge length. Some manufacturers of mono-Si wafers provide pseudo-square wafers to save material cost; a few cell manufacturers prefer 125 mm wafers. Since silicon is a brittle material, its strength is mainly limited by imperfections in the crystal lattice including lattice dislocations, grain boundaries (in poly-Si), precipitated impurities and microcracks [19]. While micro-cracks may be introduced at different process steps during wafering (as saw damage), cell or module manufacturing, the other imperfections mentioned go back to crystallization. Mechanical strength of solar cells is usually measured using either a four-point [19] or a ring–on-ring [20] bending test.

Breakage risk increases with wafer area, which is one limiting factor for increasing cell formats beyond 156 mm. Breakage risk also increases with the reduction of the wafer thickness, limiting cost saving potentials in silicon material. Theory predicts a quadratic relationship between breakage force F and wafer thickness, while some authors [20] find a linear relationship for breakage forces measured with a ring on ring breakage tester on wafers with thickness between 120 and 320 mm.

Cell thickness is expected to decrease from currently 180 µm in order to save expensive solar grade silicon material. From a functional perspective, a cell thickness of a few tens of µm would be enough to achieve reasonable cell efficiencies. Thickness reduction is mainly limited by the wafer sawing technology and by the increasing fragility of thinner cells. Current cell and module processing equipment would encounter significant production yield losses through breakage if wafer thickness would drop below 150 µm. **Kerfless wafering** technologies based on **lift-off** or on **epitaxial growth** are under development. They are able to produce very thin wafers, down to a few tens of microns.

2.8 Thermomechanical properties

In module production, materials with widely spread **coefficients of thermal expansion** (CTE) are combined. When solar cells are interconnected, silicon (CTE = 2.6×10^{-6}/°C) and copper interconnectors (CTE = 16×10^{-6}/°C) are joined at temperatures in the range of 180–220 °C, depending on the solidification temperature of the used solder material. In the course of cooling to room temperature, increasing CTE mismatch stress is partially absorbed by copper deformation. If the remaining stress on the solar cell is too high, it causes microcracks or joint breakage. During lamination, glass (CTE = 8×10^{-6}/°C) and polymer films (CTE = $50...250 \times 10^{-6}$/°C) are joined with the cells at temperatures in the range of 150 °C. When the laminate cools down to room temperature, CTE mismatch stress has to be absorbed by the polymeric layers.

In module operation, the material stack experiences cyclic temperature changes over its entire life-time of 25-30 years. Diurnal temperature cycles may be particularly demanding in desert regions with daytime module temperatures above 60 °C followed by nighttime frost.

2.9 Cell metallization and contact pads

Both the front and rear side of solar cells are usually equipped with **metallization**. Those metal structures or sheets applied on the wafer collect the generated current from the cell area as a positive cell terminal and distribute the current into the cell area as a negative terminal. The lateral conductivity of the metallization is much higher when compared to the underlying doped silicon.

For series interconnection, the positive and negative terminals of neighboring cells have to be interconnected. This connection is achieved by external metal **interconnectors**, which have to be joined to the cell metallization. The second task of the cell metallization is therefore to provide contact areas for the cell interconnectors. To allow soft-soldered or glued joints between the cell metallization and the interconnector, cells are usually equipped with enlarged, discrete, or continuous **contact pads** as part of the metallization. In the common way of speaking, the metal applied directly to the wafer by firing, plating, or fusion welding processes is denoted as cell metallization. In contrast, cell interconnectors are applied by soldering or conductive glueing on top of existing cell metallization, usually on top of dedicated contact pads. Cell interconnectors also support current collection from the cell area, depending on cell and module design, in addition to handing over current from one cell to the next cell. Cells require metallization in order to be characterized electrically in a cell flasher device, but no interconnectors.

Most solar cells on the market provide their positive and negative contact on different sides of the wafer, which means that the front side of a cell needs to be contacted to the back side of the adjacent cell for serial interconnection (fbc type, **front-to-back contact**). **Back-contact** (bc) cells provide both polarities on the rear side which reduces or completely avoids optical shading on their light receiving (active) front side [21].

2.9.1 Front-to-back contact cells

On the active cell side, metallization has to be applied intermittently to minimize cell shading. Bifacial cells have two active sides, which both need intermittent metallization. The discrete, narrow metallization lines called **fingers** only cover 3–4 % of the active cell area. Intermittent metallization requires the underlying cell layer to exhibit a sufficiently low sheet resistance (order of magnitude is 100 Ω). This is necessary to minimize series resistance losses for lateral current transport to the fingers at a finger spacing of about 1.7–1.8 mm. The sheet resistance is defined as the ratio of the volume resistivity of a uniform sheet (the emitter in this case) to its thickness (equation (2.34)):

$$r_E = \frac{\rho_E}{d_E}, \quad (2.34)$$

where

r_E emitter sheet resistance [Ω],
ρ_E emitter volume (bulk) resistivity [Ω · μm],
d_E emitter thickness [μm].

The left side of Figure 2.17 displays a front side metallization with about 90 fingers running horizontally. For convenient soldering, cells used to be equipped with continuous, rectangular contact pads on both sides, so called **busbars**. The front side busbar width was similar to the interconnector ribbon width, the rear side busbar width was even broader. To reduce silver consumption, rectangular busbars are often reduced to shaped busbars (three vertical lines on the left side of Figure 2.17) or even to separate contact pads (18 spots on the rights side of Figure 2.17). On the front side, the metallization fingers collect the current from the emitter and conduct it to the busbars.

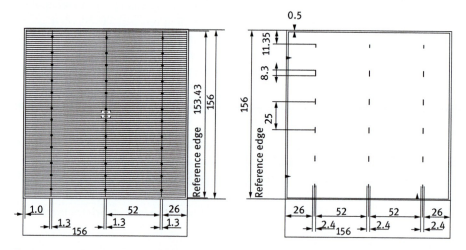

Fig. 2.17: Front and rear side metallization scheme for a 3-busbar Q6LMXP3-G3 solar cell. Reprinted with the permission of Hanwha Q-CELLS.

The number of busbars has been increased over time from two busbars to 3–6 busbars for ribbon interconnectors or up to 15 busbars for multiwire interconnectors. A larger number of busbars allows shorter cell fingers which in turn reduces the maximum current in the finger and thereby allows reduced finger cross sections. Since the width of the busbars is reduced correspondingly, neither will the shading increase, nor the demand for silver. On a 5-busbar cell, busbar and interconnector width will be chosen at or below 1 mm.

By calculating the series resistance loss in each finger and multiplying the result with the number of fingers, the total series resistance loss in the front side fingers, P_F, can be derived (equation (2.35)). The loss decreases with the second power of the number of busbars, which explains the need for increasing numbers of busbars:

$$P_F = \frac{d_F}{12 n_{BB}^2} \frac{\rho_F}{A_F} I_{cell}^2, \tag{2.35}$$

where
- P_F series resistance power loss in the fingers on one side of a solar cell,
- d_F finger distance,
- ρ_F electrical (volume) resistivity of finger material,
- I_{cell} cell current,
- n_{BB} number of busbars,
- A_F finger cross section.

Equation (2.35) also reveals that the losses increase with the second power of the total cell current. If a cell is divided in two or more cells (cutting transversely with regard to the busbars), the cell current is reduced without reducing the number of busbars. This is an appropriate measure to reduce losses. On the other hand, if cells are divided alongside the busbar, not only the current per cell, but also the number of busbars per cell is reduced accordingly. In this case, the series resistance losses remain unchanged. The most efficient approach in terms of resistance losses is to combine transverse cell division, which reduces cell current, with increasing numbers of busbars.

Front-side metallization usually consists of porous silver produced by sintering (firing) of a screen-printed silver paste. The paste also contains borosilicate glass powder with added lead oxide, which establishes the silver-silicon contact. After firing, a porous glass layer is located underneath the bulk material of the finger.

As an attempt for saving costly silver, copper has been proposed as a silver substitute for cell metallization. After ablation of the AR coating (Section 2.10) or deposition of a thin seed layer, a nickel barrier layer of about 1 µm is electroplated on the silicon, followed by about 15 µm of copper and a few 100 nm of silver on top. The electrically conductive barrier layer is required to prevent the diffusion of copper atoms into the silicon.

Silver **plating** may be used not only to coat copper fingers, but also to partially or entirely substitute screen-printed silver. Plating leads to a pure metallic, dense phase with an electric conductivity several times higher than in case of the porous sintered material. In consequence, much less of the expensive silver material is required to achieve the same finger conductivity.

For the cell interconnection process, the morphology and composition of the contact pads are critical parameters. The front side cell pads are usually processed in the same way as the cell fingers. Pastes with different silver content are used in the industry, possibly varying form 70–90 %. In contrast, plated silver is very dense and pure. Different materials and pastes are employed for the rear-side metallization. These surfaces may react differently in terms of silver diffusion when they are in contact with molten solder and in terms of stress relaxation after solder solidification. The porous metallization layers which make up the contact pads also show different diffusion properties for solder components.

If the rear cell side is inactive, shading is of no concern and metallization is applied over the entire cell area. The material mostly used for rear-side metallization of fbc cells is screen-printed and fired aluminum. The aluminum forms a highly doped p+ layer on the wafer, the **back surface field** (BSF), of about 10 μm thickness. The BSF provides moderate rear surface passivation by reflecting the minority charge carriers of the base, the electrons. This first layer is covered by a dense eutectic aluminum-silicon layer (AlSi12.2) several μm in thickness, which melts at 577 °C. There follows a porous layer of about 30 μm thickness on top. This layer consists of sintered, granular aluminum particles 5–20 μm in size, traces of silicon, and a glass matrix. The sheet resistance of a screen printed aluminum metallization is of the order of 10 mΩ.

Since aluminum is difficult to join with cell interconnectors, designated contact areas are spared and covered with silver metallization instead. The aluminum and silver metallization overlap to allow current flow to the contact pads. Silver paste designed for rear-side application usually contains small amounts of aluminum to improve passivation underneath the contact pads. These rear side contact pads may be implemented as continuous busbars or as discrete pads. Front-side and rear-side busbars (or contact pads) are located on corresponding places on both sides of the solar cells. In this way, soldering tools can simultaneously create solder joints on both sides of the cell, and the mechanical pressure required for heat conduction and ribbon fixation only exerts limited stress on the fragile cell.

For further silver reduction, some cell designs completely omit contact pads. Ribbons or interconnection wires are directly joined to the front-side finger metallization by adapted soldering or glueing processes. On the rear side, it has been proposed to apply a **tin pad** directly on the screen printed aluminum layer [22] in order to reduce silver consumption. Aluminum instantaneously forms a very stable Al_2O_3 oxide layer a few nm in thickness when exposed to air, displaying a melting point of about 2000 °C. Aluminum surfaces are therefore not compatible with common soft-soldering materials. The wetting of the aluminum with tin is therefore achieved by a sonotrode that oscillates at ultrasonic frequency. Once the tin pad has been applied on the solar cells, they can be processed similarly to silver pad cells. Figure 2.18 shows a solder layer applied on top of the bulk aluminum layer.

In order to improve the rear-side passivation and to increase its reflectance for infrared radiation, **dielectric rear-side passivation** replaces the full area aluminum BSF. On top of the dielectric layer (e.g. SiO_2), aluminum is either evaporated or applied as a sheet, and local contacts are established to the wafer via the dielectric layer.

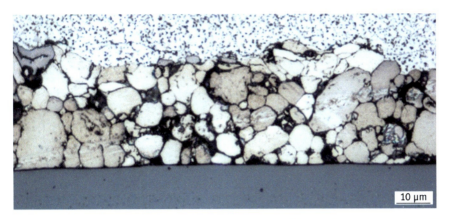

Fig. 2.18: Optical microscope image of a cross section prepared by metallography showing solder on top, porous aluminum metallization inbetween and the grey silicon wafer at the bottom (Fraunhofer ISE).

2.9.2 Back contact cells

Two different concepts are used in commercial back contact cell technology. The **metal wrap through** (MWT) solar cell architecture with its front side emitter is quite similar to common fbc cells. The major difference lies in the absence of front-side contact pads. Instead, holes are opened in the wafer by a laser, their border is doped together with the emitter, and the hole is filled with metallization paste. The cell current flows from the emitter through the fingers and the holes to the rear side of the cell, where additional contact pads for the emitter polarity are provided. The left side of Figure 2.19 shows the front side of a commercial MWT cell with four busbars and four broadened spots on each busbar. These spots are not intended for contact pads; they constitute the top view of holes filled with silver. Their rear-side counter parts are visible on the right side of Figure 2.19 as central parts of 16 rings. The rings separate emitter and base polarity. The 15 simple circles correspond to the base polarity. The number of contact pads, totaling 31 in this particular case, are distributed over the entire cell area. Cell interconnectors need to reach these **distributed contacts** and lead the current to the cell edges.

As for **interdigitated back contact** (IBC) or, more precisely, **back junction back contact** (BJBC) cells, the emitter has been completely moved to the cell rear side, and the front side looks homogeneously black. The full area emitter has been reduced to a set of strips which bear the emitter metallization. The base metallization is arranged in between the emitter strips, leading to an interdigitated busbar architecture. BJBC cells are usually provided with **edge contacts** (right side of Figure 2.20), meaning that the cell metallization conducts the cell current to opposite cell edges and interconnectors only have to span the cell gaps. At the moment, busbar conductivity limits cell formats

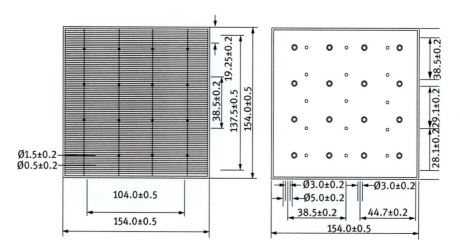

Fig. 2.19: Front- and rear-side metallization scheme for the JACP6WR-0 MWT poly-Si solar cell from JA Solar.

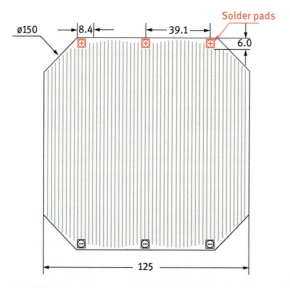

Fig. 2.20: Front view and rear side metallization of a Sunpower A-300 BJBC cell.

to 125 cm. Distributed contacts on BJBC cells would introduce low-efficency-areas due to prolonged paths for charge separation.

2.10 Antireflective texturing and coating

The **refractive index** n of a material denotes the factor by which the speed and consequently the wavelength of light are reduced when compared to propagation in vacuum. This index also controls the splitting of light, incident on a material interface at a given angle of incidence, into a reflected and a transmitted fraction.

The refractive index of silicon varies between 3.5 and almost 7 in the wavelength range from 350 to 1100 nm [23]. Due to the high mean refractive index of silicon of about 3.9, a plane air/silicon interface displays an effective reflectance of 36 % for normally incident light. To improve the light penetration into the wafer, in the process of solar cell manufacturing the wafer surface is both textured and coated. It has to be noted that the effect of **antireflective** (AR) measures on cell level is modified by cell encapsulation.

Texturing (Fig. 2.21) aims at generating two or more reflection processes for normally incident rays, instead of a single reflection, thus increasing the effective transmittance of the surface and "trapping" the light. A second valuable effect of texturing is the prolongation of the ray paths in the wafer, which increases the light absorbtance of the wafer in the infrared spectral range. Different texturing processes are applied, depending on the wafer material.

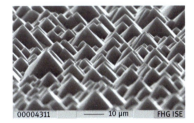 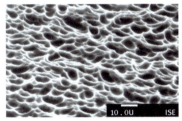

Fig. 2.21: SEM image of anisotropic texture obtained by alkaline etching on mono-Si wafer (left) and isotropic texture by acid etching on poly-Si wafer (right) [26].

On mono-Si material, alkaline etching leads to the formation of **random pyramids** on the wafer surface. Pyramid edge length varies from the submicron range to several microns. All pyramid facets are inclined by 54.74° with respect to the cell plane. Light incident normally to the cell plane therefore only encounters two classes of paths in an idealized pyramid landscape: 89 % of the light is reflected twice at 54.74° and 15.79° local incidence angle, and the rest is reflected three times, at 54.74°, 15.79°, and 86.32° [24]. Some flat regions without pyramids may persist after etching, which then strongly contribute to residual wafer reflectance. Anisotropic alkaline texturization can reduce the effective reflectance to about 12 %, which is roughly one third of the plane surface reflectance.

On poly-Si material, acid wet chemical etching in HF is commonly applied to obtain an **isotropic texture**. The effective surface reflectance can only be reduced to about 26–29 % [25], which implies substantially higher reflection losses compared to textured mono material. In consequence, poly-Si solar cell efficiency relies strongly on further measures like coating and module encapsulation.

Several AR processes have been suggested to provide improved textures especially on poly-Si material, e.g. (dry) plasma texturing, honeycomb texturing, and reactive ion etching (Figure 2.22). Honeycomb texturing using photolithography, nanoimprint lithography, or inkjet masking has been proven to reduce surface reflectance to about 6 % [27] on lab samples, yet it requires additional process steps for masking and stripping. Reactive ion etching has also been reported to reduce surface reflectance below 20 % on lab samples by producing needle-like nanoscale surface structures without the need for masking. The needle structure additionally scatters incoming rays, thus extending the ray paths in the wafer and improving absorptance.

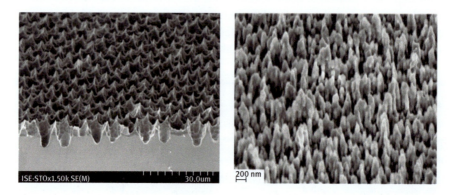

Fig. 2.22: SEM images of honeycomb texture on poly-Si wafer by nanoimprint lithography [28] (left) and by reactive ion etching (right, [29]).

Figure 2.23 shows measured spectral reflectance of common textures on poly-Si and mono-Si wafers, compared to a honeycomb texture on a poly-Si wafer. The latter achieves an effective reflectance of 6 %.

To further reduce reflectance, **AR coatings** are applied to the silicon surface. With their intermediate refractive index, they provide a smoother transition between the low refractive index of the environment and the silicon wafer with its relatively high index. For an encapsulant with an index of 1.5 and silicon with 3.9, the reflectance of a flat interface amounting to 20 % could be reduced by a single AR-layer with a refractive index of 2.4 to 11 %. Yet, this result is not sufficient for solar cells. The AR effect of the layer can be further improved for a defined wavelength interval by choosing a layer thickness close to one quarter of the specified wavelength. The most common solution for solar cell AR and passivation treatment is a single silicon nitride (SiN_x) layer

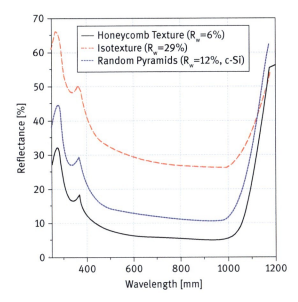

Fig. 2.23: Measured spectral reflectance of differently textured wafers, before AR coating [27]

with an intermediate refractive index of about 2.1 and a thickness in the range of 70–80 nm. Increasing the refractive index of SiN_x towards the ideal value is not possible without raising absorbtance. For achieving highest cell efficiencies, double-layer AR coatings are used (e.g. SiN_x together with SiO_xN_y). Physical concepts underlying AR layers are discussed in Section 3.2.1. It has to be noted that the stack air/coating/silicon has different optical properties than the stack encapsulant/coating/silicon due to the refractive index change in the first medium. AR coatings should therefore be optimized with respect to the module environment which usually implies optical contact to an encapsulant.

3 Module design, materials, and production

Wafer-based PV modules comprise one or more solar cells that are connected and encapsulated according to their application requirements. Since a single solar cell only provides voltages of around 0.5 V at its maximum power point, usually several solar cells are connected in series. Most modules intended for grid connection consist of 60 cells with 156 mm edge length, arranged in 6 strings of 10 cells each (Figure 3.1). All 60 cells are connected in series and two additional contacts are provided to allow placement of bypass diodes that bridge each of the three 20-cell-substrings. Therefore, four contacts in total are fed through the perforated backsheet (dark rectangle at the bottom) to be connected to the junction box. The box on the rear side of the module is not visible in this top view. Both views in Figure 3.1 show the glass pane on top, followed by the front side encapsulant layer, the cell matrix, the rear side encapsulant layer, and the backsheet.

Common 60-cell modules are 990 to 1000 mm wide and 1640 to 1680 mm long. The height is determined by the frame and typically reaches 33–50 mm. Some manufacturers use cells with 125 mm edge length and place 72 cells in one module.

An intermediate adhesive tape or sealant is applied in between the frame and the laminate edge, illustrated as black layer in the cross-sectional view. Towards the glass edge, the encapsulant layer becomes very thin. During lamination, the backsheet is pressed towards the glass edge by the laminator membrane, displacing the encapsulant and improving the module edge tightness.

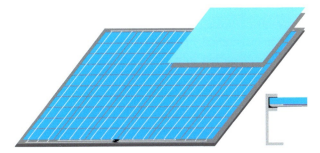

Fig. 3.1: Drawing of a 60 cell PV module assembled with 3 busbar cells, the cross sectional view on the right includes the module frame.

Figure 3.2 shows the bottom left corner of the same module with the cell interconnector pathway from the front side of each cell to the rear side of the adjacent cell and the string interconnection scheme.

The cell matrix is highly fragile and challenging to handle. It requires cover materials for protection and stabilization as well as intermediate layers between cells and covers that act as encapsulant.

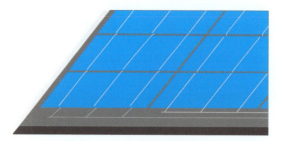

Fig. 3.2: Detail of module corner highlighting the cell and string interconnection scheme.

The front cover, usually a glass pane, supports the cell matrix. It conducts mechanical loads into the mounting structure, usually via the module frame. The frame stabilizes the module against mechanical loads of static or dynamic nature like snow or wind. The front cover needs to protect the cells against hail impact. The rear cover, which may consist of a polymer sheet or a second glass pane, also serves as a mechanical protection. Both covers act as diffusion barriers, especially against moisture ingress. The covers also need to resist dielectric breakdown and insulate against leakage currents, since the cell matrix may operate at a voltage of 1000 V or more vs ground. The front cover and the front side encapsulant layer need to be highly transparent for solar radiation in the spectral range between 350–1100 nm which is relevant for PV conversion.

The junction box usually hosts bypass diodes. They reduce the risks of excessive performance loss due to partial shading and protect the module against damaging hot-spots. The box may also integrate active electronics like power optimizers or microinverters.

Cables and plugs conduct currents of typically 8–9 A. As all external parts of the module, they need to be designed to resist ambient conditions over the service life. Warranties may reach up to 30 years of operation with a final nominal module power loss below 20 %.

The entire module design may be challenged by very high temperatures of 80–90 °C, very low temperatures, and numerous temperature cycles, depending on ambient conditions and particular installation. During its service life, it will experience severe doses of UV radiation, e.g. in the range of 5 GJ/m^2 (1500 kWh/m^2) in 20 years of exposure in a moderate European climate [30]. Elevated temperatures, humidity, and UV impose serious restrictions on the choice of polymers to be used as encapsulant and cover materials.

Figure 3.3 shows a cross section of the module. Glass thickness is chosen to match mechanical load specifications, which usually requires 3.2–4 mm glass. Each encapsulant layer (2), (4) has a thickness of 0.4–0.5 mm. Cell spacing is typically 2–3 mm. The interconnector ribbon (6) cross-section plane is orthogonal to the current flow direction. Full cell width is usually 156 mm with 3–5 equally spaced ribbons. Backsheet thickness is a few hundred microns, depending on the system voltage specification.

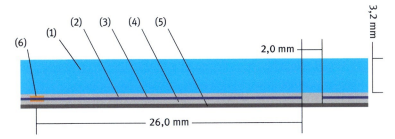

Fig. 3.3: Cross section view of a PV module with 3.2 mm front glass (1) on top, front (2), and rear (4) encapsulant, cell matrix (3), backsheet (5) and interconnection ribbon (6).

3.1 Cell interconnection

3.1.1 Ribbons and wires

Solar cells are serially interconnected to **cell strings** in order to deliver electrical power at moderate current and convenient voltage levels. Since series resistance losses in connections increase with the second power of the current, low current design is preferable. A 20 % efficient solar cell sized 156 mm × 156 mm may deliver 8–9 A of current at mpp. Additionally, inverters tend to operate more efficiently if their DC input voltage exceeds the AC output voltage required for grid connection.

Serial cell interconnection requires the cell current to be collected over the cell area, to be transported to the cell edge, to pass a cell gap and reach the opposite polarity of the neighbouring cell, where it needs to be redistributed into the cell area (Figure 3.4).

Fig. 3.4: Schematic current flow from interconnection wire (long arrows showing large currents) into cell fingers and cell active area (smallest arrows) on the top ("sunny", emitter, negative) side of a p-type solar cell in conventional notation.

In common c-Si modules, the cells are interconnected using copper flat wires called **ribbons**. They consist of soft temper copper bearing a solder coating of 10–25 µm thickness. Cell soldering processes usually rely on the solder from the ribbon coating and do not introduce any additional solder material. For quality assurance it is very important to ensure a uniform solder coating thickness on both sides of the ribbon.

Figure 3.5 shows stress-strain curves of ribbons with different material quality. The left figure uses a magnified horizontal axis to highlight the elastic regime with strain proportional to stress. The intersection of the dotted line starting at 0.2 % strain and having the same slope as the elastic section of curve 1 with the curve itself leads to a characteristic stress value on the vertical axis, the $R_{P0.2}$-value. This value is called the **offset yield strength** (or elastic limit), since it is difficult to identify a distinct transition limit from elastic to plastic deformation in ductile materials like copper. Ribbon 1 displays an $R_{P0.2}$-value around 140 MPa (or N/mm^2). A soft material quality (ribbon 3), which is favorable for low-stress interconnections, is achieved with highly pure copper material and special thermal treatment for annealing. Soft ribbons provide $R_{P0.2}$-values in the range of 70–100 MPa. Strain that usually occurs in ribbon handling may increase the yield strength of the copper due to work hardening effects.

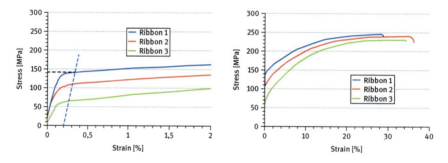

Fig. 3.5: Stress-strain curves of three different ribbons.

A second important mechanical parameter is the elongation at fracture or fracture strain. Copper displays ductile fracture with necking after plastic deformation. High fracture strain above 25 % is favorable for cell interconnection, since particularly in the gaps between solar cells ribbons experience severe strain in module operation.

In PV as in other electric applications, copper qualities of high purity are required for minimizing series resistance losses. Copper in Cu-OFE (oxygen-free electronics) quality shows a purity above 99.99 % mass and provides an electric conductivity between 58–59 MS/m at 20 °C. Cu-ETP (electrolytic tough pitch) quality with a purity above 99.9 % mass provides 57 MS/m conductivity.

The copper ribbon is either rolled from a coated round wire or it is cut from a solder-coated copper sheet. Typical rectangular ribbon cross sections for 3-busbar cells amount to 1.5 mm width and 150–200 µm height. On 5-busbar cells, ribbons are somewhat less than 1 mm in width. Due to their width, ribbons typically cover approx. 3 % of the cell area. The width of the ribbons is constrained by shading losses. The ribbon height is constrained by the thickness of the encapsulation layer and by issues related to the stability of the joints on the solar cell. These constraints require a compromise with regard to the total cross section and associated series resistance losses in the ribbon. These electrical losses typically amount to 3–4 %.

Aluminum is regarded as a potential substitute for copper in cell interconnection. Its electric conductivity is one third less then copper, but it is more than two thirds cheaper. Yet, aluminum raises some challenges related to joint formation, increased mismatch with regard to thermal expansion, and reduced ductility.

Series resistance losses also occur in the cell fingers. In order to reduce these losses, the industry moved from two busbars per cell to three and more. By increasing the number of ribbons (or wires) per cell, the current path in the fingers is shortened, and maximum current values are reduced.

Common ribbons have a flat, glossy surface. Incident light is specularly reflected, according to the coating's reflectivity, preserving the incidence angle. Typically, almost half of the incident light is absorbed while the rest is reflected back and leaves the module (Figure 3.6, top left). This means that 100 % of the light incident on the ribbon is lost.

Different surface treatments of the ribbon have been proposed to reduce optical losses.

A saw-tooth profile [31] together with a glossy surface reflects normally incident light at angles that exceed the total internal reflection (TIR) angle (Figure 3.6, top right; red ray is totally reflected). The decisive angle is the TIR angle of the encapsulant with respect to air, even if a cover material with a different index of refraction separates the

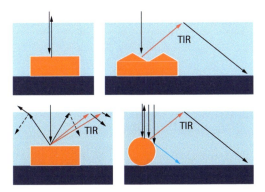

Fig. 3.6: Schematic light paths for different interconnector cross sections: common ribbon with specular symmetric reflection (top left), saw-tooth profiled ribbon with specular nonsymmetric reflection (top right), scattering ribbon with diffuse reflection (bottom left), and round wire with specular divergent reflection (bottom right) in a laminated module showing for normally incident light.

encapsulant from air. Depending on the refractive index of the encapsulant, the TIR angle varies between 42 and 45°.

Most of the secondary reflected rays that encounter TIR will reach the active cell area and contribute to current generation. A glossy silver coating can additionally reduce the absorption losses on the ribbon surface down to values in the range of 5 %. The saw-tooth approach works best for light incidence close to the orthogonal direction. Instead of embossing the structure into the ribbon, a film with the same functional surface structure and highly reflective coating may be applied on a flat ribbon [32].

A different approach for improving optical ribbon performance uses a white, diffusely scattering, high reflectivity coating on the ribbon surface. Backscattered light will be partially reflected at the glass/air interface, especially those rays that encounter TIR (Figure 3.6, bottom left; red rays will be totally reflected). Since the ribbon needs to be joined to the front and rear side of neighboring cells, the coating is applied on one side of the ribbon intermittently, only over the length of each second cell. The coated segments are soldered to the cell front side and the uncoated segments to the cell rear side.

Using wires with a round cross section and glossy, high reflectivity coating instead of flat ribbons also proves optically advantageous (bottom right in Figure 3.6). Light incident close to the perimeter can be directly reflected to the active cell area. In the adjacent section of the wire, reflected rays point to the glass/air interface and are redirected to the cell by TIR (Figure 3.6, bottom right; red ray is totally reflected). With 10–20 wires distributed over the width of the solar cell, the maximum current and the current path length in the cell metallization fingers is substantially reduced. Figure 3.7 shows a ribbon-based interconnection with three busbars and a multiwire interconnection. Cell metallization fingers are not displayed.

Fig. 3.7: Drawing of a solar cell string connected with ribbons on 3 busbars (left) and with 12 wires (right).

Figure 3.8 shows a detail of a multiwire module using 15 wires. The solar cell metallization requires no busbars, but the fingers have been adapted to provide enlarged areas for the wire joints. These solder pads improve joint stability.

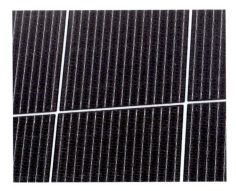

Fig. 3.8: Detail of a multi wire module prototype. Reprinted with the permission of SCHMID Group, SCHMID Technology Systems GmbH.

3.1.2 Structured interconnectors

In common cell designs, interconnectors collect current over the length of the solar cell, leading to a linearly increasing current in the interconnector from one cell edge towards the opposite cell edge (Figure 3.2). In ribbons with constant cross section, the current density therefore varies over the cell.

For a given volume of copper, the most efficient material use could be achieved if the current density in the interconnector is kept constant (left side of Figure 3.9). This is due to the dependence of the series resistance losses on the second order of the current density. Constant current density would require a tapered interconnector design

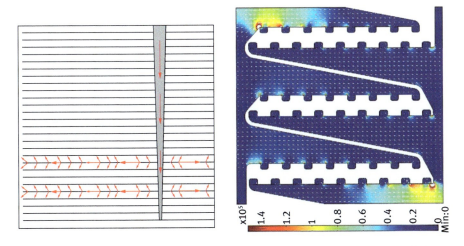

Fig. 3.9: Schematic current flow from tapered interconnection wire into cell (left), FEM calculated power loss density (arbitrary units) in a tapered interconnector for back contact cells (right, image from [33]).

where the cross section relative to the current direction also increases linearly, equally to the current. Since production and handling of such tapered ribbons is costly, they are not currently used for front-to-back contact cells.

Tapered designs have been proposed for the rear-side connection of cells with distributed contacts, where connectors do not run into any conflict with cell shading. The right side of Figure 3.9 shows a design for both cell polarities with cross sections adapted to local current intensity. Bottlenecks appear in yellow and red, indicating increased power dissipation. The structured interconnector features a broad, tapered main bar, that conducts the increasing current, and terminals that are linked to the main bar. The small cross section of the terminals is useful to relieve thermo-mechanical stress from the joints. The stress originates from temperature variations during module production (especially soldering) and from module operation. Figure 3.10 displays a prototype MWT (Section 2.9) string with two cells and nontapered interconnectors with structured terminals. The green color originates from an insulating lacquer that prevents short circuit between the base polarity underneath and the emitter polarity connected to the structured interconnector on top of the laquer.

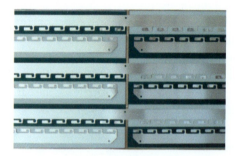

Fig. 3.10: Detail of an MWT cell string showing structured interconnectors with broad main bars for increasingly larger currents and small terminals which are soldered to the cell contact pads.

Back contact cells may be equipped with a reinforced metallization, usually based on copper, which transports the entire cell current to opposite cell edges. In this edge contact design, interconnectors only need to transport current over the narrow cell gaps (top left of Figure 3.11). A straight ribbon (bottom left of Figure 3.11) provides insufficient flexibility and would lead to joint breakage under mechanical or thermomechanical loads. A bow shape offers more stress relief for relative cell displacements in string direction (bottom center of Figure 3.11), but it may still be too stiff. If the bow is interrupted into two or multiple bows, the stiffness is drastically reduced at marginal conductivity losses (bottom left of Figure 3.11).

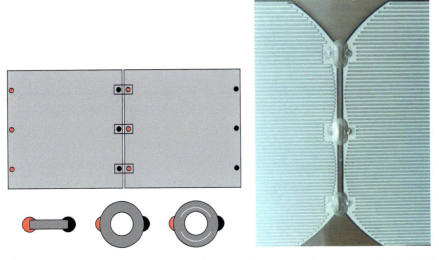

Fig. 3.11: Edge contact arrangement on a BJBC cell (top left), schematic interconnector details for BJBC cells showing a direct, single bow and multiple bow connection (bottom left) and detail of a BJBC string with Sunpower A-300 cells showing multiple bow interconnectors.

3.1.3 Conductive backsheet

For back contact module assembly, a structured conductive layer may be created by first laminating a metal sheet to a module backsheet and structuring the metal layer according to the desired interconnection scheme. These structures may be obtained by mask printing, etching and stripping, similar to the printed circuit board (PCB) process. As metal sheet, 35 µm thick copper is in use, or sandwiches consisting of copper and aluminum. In order to reduce cost, laser and milling processes have been suggested for structuring as an alternative to etching (Fig. 3.12).

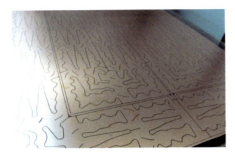

Fig. 3.12: Laser-structured conductive backsheet for full size module. Reprinted with the permission of Eppstein Technologies.

Module designs using back-contact cells and structured conductive backsheets require an intermediate insulating layer. Screen-printed insulating finish is applied either on the rear side of the cells or on the conductive layer.

3.1.4 Cell shingling

In a shingle-type string, neighboring cells overlap such that they can directly transfer current to the neighboring cell in the string without any additional interconnector (Fig. 3.13). The front and rear side contact pads are arranged at opposite cell edges. The joints inbetween cells will usually require electrically conductive adhesives (Section 3.1.7) as opposed to solder in order to provide the necessary flexibility.

Fig. 3.13: Drawing of a shingled cell string.

Shingling avoids optical losses stemming from the inactive area of the front side contact pad, since it is covered by the active area of the neighboring cell. It also omits inactive module area commonly required for cell gaps. Because no interconnectors are present to support current collection from the cell area in string direction, the respective path length is restricted by the conductivity of the cell metallization. Usually, shingling will be used for cells with reduced length in string direction (e.g. quarter cells). Shingling largely avoids inactive module area as well as inactive cell area, thus enabling the assembly of modules with high power and efficiency. Cell shingling has been used for device-integrated modules where maximum efficiency is required due to the restricted module area. Recently, cell shingling has also been suggested for large PV modules and a module power of 400 W of has been reported using the equivalent of 72 cells.

3.1.5 Soldering processes

In PV applications, soldering establishes an electrically conductive und mechanically stable joint in between metal parts by means of an additional metal with a lower melting point, the solder. In the process of soldering, all parts are heated above the solder melting temperature, without reaching the melting point of the parts to join. The molten solder spreads over the solid surfaces, and subsequent cooling results in a solid joint.

Sufficient **wetting** is a precondition for stable solder joints. Wetting can be quantified by measuring the **contact angle** (Figure 3.14) in between the flat substrate and the liquid solder surface at the point where the solder-gas interface touches the substrate [34]. Wetting measurements are performed in an inert atmosphere to prevent surface oxidation.

Poor wetting leads to drops that approach a spherical shape at contact angles larger than 90° (right side of Figure 3.14). Flat drops with contact angles in the range of a few tens of degrees indicate proper wetting. The contact angle depends on the

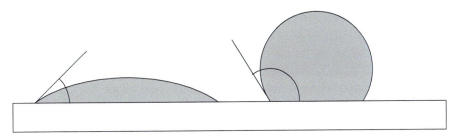

Fig. 3.14: Liquid drops on a substrate with small and large wetting angle.

balance of the **surface tensions** of the involved material interfaces. Surface tension opposes the enlargement of the interface area. It is defined as the work required to increase the interfacial area, with the unit N/m. When contact angles are measured on solar cells, diffusion and oxidation effects need to be considered. For example, diffusion may deplete the cell contact pad surface of silver in the course of the measurement and **dewetting** may occur. Contact angle measurements are also influenced by surface roughness, temperature, time, and hysteresis effects. As an alternative to direct contact angle measurement, wetting can also be measured by a wetting force balance. The balance measures the force needed to extract a specimen from a bath of molten solder and requires some effort to prepare special wafer samples.

In most soldering applications, including common solar cell interconnection, **fluxes** are used to remove oxidized layers from the interfaces and to improve and accelerate the wetting of the parts to be joined. Storage time and conditions of the cells and ribbons affect the persistence of these layers and thereby the amount of required flux. Fluxes are usually composed of activators, vehicles, solvents, and possibly additives. The **activator** is the reactive component that attacks and dissolves oxidized layers. For cell soldering, weak **organic acids** like adipic acid are used as activators [35]. Since no subsequent cleaning of residual flux can be performed in module production, only so-called "**no-clean**", noncorrosive fluxes are acceptable. The vehicle is intended to flush away reaction products and to temporarily protect the exposed metal surface from reoxidation. Solar cell soldering is performed under atmospheric conditions, where elevated process temperature accelerates the formation of a new oxide layer. Alcohols are often used as solvent, primarily isopropyl alcohol. Flux is only applied in very small quantities in string manufacturing. So-called "**low solid**" fluxes are used with a solid content below 2 % (weight).

After flux application to the solder coated ribbon or the cell contact pad, the solvent is completely evaporated at temperatures of about 80 °C, leaving dry surfaces with small crystal grains of organic acid. In this solid state, the organic acids are inactive.

For the interconnection of bc cells, it is also possible to use **solder paste**. The paste can be applied by dispensing or by screen printing, either on the solar cell or on the interconnector. When solder paste is used, the structured interconnectors or conductive backsheets only require a thin layer of tin or other materials to protect the surface against oxidation. The solder paste contains solder particles and substantial amounts of flux. Solder pastes are difficult to use on the active cell side, since flux leakage during soldering may contaminate the cell.

Various heating technologies for solar cell soldering are being used, with or without contact heat transfer (Section 3.4.2). The cell soldering process consists of three basic phases: heating, maintaining and cooling. During the heating phase, the melting point of the organic acid is reached first (e.g. adipic acid at 151 °C). With melting, the organic acids are activated and disrupt metal oxide layers. With further increasing temperature, the melting point or melting range of the solder is reached. The mechanical pressure acting on the interconnector for fixation or heat transfer purposes will cause some local displacement within the liquid solder layer. Additionally, molten solder will spread over wettable surfaces, especially in narrow gaps where it is driven by capillary action (Fig. 3.15).

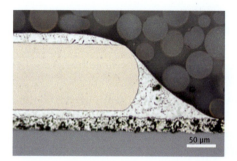

Fig. 3.15: Metallographic image showing a cross-sectional view of a copper ribbon soldered on a screen printed contact pad of a solar cell on the bottom [38].

After wetting took place, the elevated temperature is maintained briefly to allow **metallic interdiffusion** as a prerequisite for a stable solder joint. For soldering on screen-printed solar cell metallization, the interdiffusion process requires a narrow control of its temperature profile in time. If the molten state is maintained too long, a depletion of silver on the contact pad surface may occur, leading to a reduced strength of the solder joint.

The cooling phase is very critical for the relaxation of thermo-mechanical stress caused by a mismatch in coefficients of thermal expansions (CTE, Section 2.8). Cooling too quickly for a given set of materials and parameters can lead to cell crack formation. The cracks may interrupt cell fingers and reduce cell power [36, 37]. Highly critical for power loss are soldering induced cracks running parallel to the busbars, since they may completely disconnect peripheral cell areas [39].

Different soldering approaches have been suggested to connect an array of wires to the cell metallization, instead of ribbons. In the US patent no. 20050241692, Leonid and George Rubin describe an electrode that consists of a transparent, multilayer film and an array of wires embedded into an adhesive layer on said film [40, 41]. This electrode is prepared such that the adhesive layer has alternating orientations pointing upwards or downwards. In the next step, the electrode is cut to the length of a two-cell-string and applied accordingly on top of a solar cell and on the bottom of the neighboring cell, with the wire array (about 18 parallel wires) always facing the cells. The adhesive is heated to achieve a preliminary bonding to the cell surfaces. The wires get in touch with the cell metallization, but the temperature is not sufficient to melt the solder. Only in the subsequent module lamination process is the solder alloy melted at temperatures of about 150 °C and solder joints are formed. The process flow requires a low melting solder as wires coating with alloy such as indium-tin or bismuth-tin (Figure 3.16). For this film electrode process, solar cell metallization only uses continuous

Fig. 3.16: Cell stringing concept using electrode with wires attached to transparent film (commercial name is SmartWire Connection Technology, reprinted with the permission of Meyer Burger.

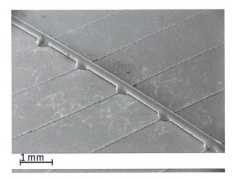

Fig. 3.17: SEM image of wire soldered to cell metallization fingers (reprinted with the permission of SCHMID Group, SCHMID Technology Systems GmbH).

fingers without any designated contact pads or busbars. Consequently, there are no critical alignment requirements between electrode and cell.

In a different approach, the array of wires is applied and soldered to the solar cell metallization (Figure 3.17) by tools resembling conventional ribbon stringers. To obtain stronger solder joints, the narrow cell fingers are locally widened to provide tiny contact pads. The wire cell strings are then processed similarly to conventional ribbon cell strings.

3.1.6 Solders

The joints between the solar cell contact pads and the cell interconnector are commonly established by soldering. The close contact at high temperature allows atoms from the solid metals (copper, silver) to diffuse into the molten solder (tin or tin-based alloys) at a high rate. Their former places can be taken by solder atoms. The resulting **alloy**, a mixture of metals, displays a concentration profile that reaches from pure component A, e.g. the solder, to pure component B, e.g. silver. In between the pure components, the concentration of component A falls from 100 % to 0 % and vice versa. The soldering temperature profile during the molten phase, especially the maximum temperature and its duration, affect the component profile and is thereby a critical parameter for the process quality assurance.

In the cooling phase, the solder solidifies and largely conserves the composition distribution reached in the molten state. Yet, **solid state diffusion** keeps altering the microstructure at low diffusion rates even at room temperature and will usually coarsen the microstructure. The diffusion rates are accelerated by elevated operation temperatures and by severe mechanical stress. One possible outcome of diffusion is the age-softening of tin-lead solder, where the precipitation of tin during storage at room temperature reduces its initial hardness achieved immediately after solidification.

In the solid state, the alloy forms crystal structures, depending on the solubility. If the size of the involved atoms and their crystal structures are very similar, e.g. for the pair copper-nickel, perfect solubility is achieved. In this case, a homogeneous single phase solid will result at any composition. It consists of mixed crystal grains (**crystallites**) with the same structure as the pure constituents. In the course of cooling, the molten alloy first turns into a paste-like condition with solid grains of mixed crystals in the residual melting. This first turning point is called the **liquidus temperature**. With further cooling, the crystal fraction grows until the alloy becomes completely solid at the **solidus temperature**. In the opposite extreme, if the species are very different, they totally demix during crystallization to form their own crystal grains (A, B). No mixed crystals can be observed in this case.

In the intermediate case of limited solubility, the species can only constitute mixed crystals within a certain range of compositions, and these mixed crystals may only be stable within a certain range of temperatures. The mixed crystals may be of the same crystal type as the pure component crystals (α) or (β), or of a different crystal type (γ). Compositions beyond the allowed mixing range will lead to the formation of different types of mixed crystals in the solid alloy, a mixture of mixed crystals so to say. There is a certain composition, the **eutectic**, where both species crystallize simultaneously and form a particularly fine-grained solid. Only in this composition a distinct **melting temperature** (or melting point) can be determined. Outside of the eutectic composition, the molten alloy first turns into a paste-like condition where solid grains of the over-eutectic species form in the melting. With further cooling, the residual melting is increasingly depleted of the over-eutectic species until it reaches the eutectic composition at the solidus temperature. At this point, both species crystallize from the residual melting and the entire material becomes solid. Non-eutectic solders therefore exhibit a melting temperature interval from solidus to liquidus instead of a melting point. The restrictions in building mixed crystals originate from differences in atom size and crystal system of the involved species, as is the case of tin-lead and tin-copper alloys.

Phase diagrams are used to map the phases of an alloy in thermodynamic equilibrium, depending on the metallic species and on temperature. In the SnPb phase diagram (top left of Figure 3.18), (Sn) denotes a phase with the tetragonal crystal structure β of (white) tin. It contains up to a few percent of dissolved lead. The maximum fraction of solvable lead decreases with cooling from the eutectic temperature. This will result in oversaturation followed by gradual demixing and lead segregation through solid state diffusion below a specific transition temperature. (Pb) stands for a phase that exhibits the face-centered cubic crystal structure α of lead. This phase is either pure lead or a solid solution of moderate fractions of Sn in lead. At the eutectic temperature of 183 °C, maximum solubility of tin in solid lead (19 % weight) and lead in solid tin (2.5 % weight) is reached. In the central part of the diagram, below the eutectic temperature, the solder will contain both (Sn) and (Pb) as distinct phases with very small amounts of dissolved species in the foreign crystallites. The SnBi phase diagram (top right of Figure 3.18) displays a lower eutectic temperature than SnPb. The maximum solubility for Sn in solid bismuth only amounts to 0.1 %.

If the alloy adopts a new crystal type γ, different from the types α and β of the pure components, the corresponding material is called an **intermetallic phase**. In general, intermetallic phases allow a variation of the component fractions within certain limits. If the new crystal structure only allows a fixed ratio of constituents as in chemical compounds with defined stoichiometry, the intermetallic phase is called **intermetallic compound** (IMC).

Common IMCs in solder joints involving tin-based solder-coated copper interconnectors on silver metallization are Cu_3Sn (ε-phase) adjacent to the copper, Cu_6Sn_5 (η-phase) adjacent to the solder, and Ag_3Sn (ε-phase) adjacent to the cell metallization. IMCs are harder and more brittle than the bulk solder. The fraction of IMCs can quickly grow at elevated temperatures. IMCs can cause fracture under stress, e.g. un-

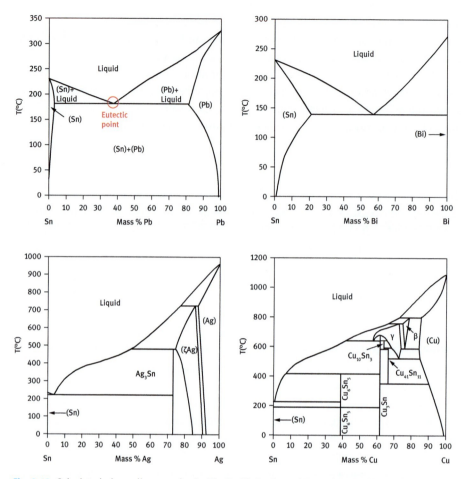

Fig. 3.18: Calculated phase diagrams for Sn-Pb, Sn-Bi, Sn-Cu and Sn-Ag binary alloys. Reprinted with the permission of the National Institute of Standards and Technology (NIST).

der temperature cycling, mechanical shock and vibrations. Solder composition, solder layer thickness and soldering process parameters have to be carefully adjusted to control the total fraction of IMCs and their distribution within the joint interfaces.

Most PV manufacturers still use eutectic (or close to eutectic) SnPb37-based solder for cost reasons and because of its convenient melting (or solidus) temperature at 183 °C. The lead addition not only reduces the melting point of pure tin, but also reduces its surface tension, thus improving wetting properties. SnPb37 solder consists of a tin-rich (Sn) and a lead-rich (Pb) phase in its solid state.

The microstructure of the bulk solder is determined by its thermal history, namely the cooling and storage parameters. Rapid cooling will produce small grains, since diffusion is stopped in an early stage. Slow cooling and long storage, especially when combined with elevated temperature, will stimulate the growth of grains or lamellar structures. Figure 3.19 shows solder joint cross sections after metallographic preparation in different segregation states, with pronounced Ostwald ripening on the right side. The silicon wafer is located at the bottom, followed by the silver metallization, and the solder layer and the copper ribbon are on top.

Silver is added to produce eutectic SnPb36Ag2 alloy, which slows down the diffusion of silver during soldering, to prevent possible silver depletion on the cell metallization surface. Silver addition also slightly reduces the melting temperature to 179 °C.

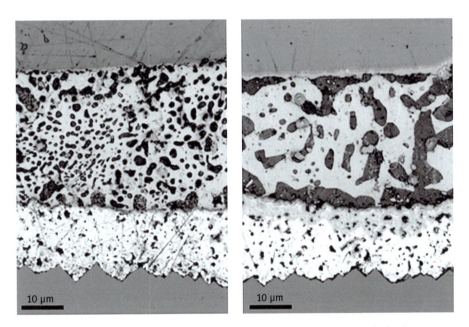

Fig. 3.19: Optical microscope images of SnPb36Ag2 solder in the initial state and after heat storage at 130 °C for 155 hours showing pronounced microstructure coarsening (Fraunhofer ISE).

3.1.6.1 Lead-free solders

Concerns about lead pollution have triggered extensive activities to replace lead. The US Environmental Protection Agency has listed lead as one out of 17 substances that pose the most severe threats to humans and to the environment. The EU directive RoHS on the restriction of the use of certain hazardous substances in electrical and electronic equipment was adopted in February 2003 and took effect in 2006. This directive restricts the use of lead and is expected to extend to PV modules in the near future.

Pure tin is not considered to be an alternative, since it is not stable below 13 °C, where it transforms into brittle "gray tin". Lead-free tin-based solders with silver addition [42] avoid this transition, but still require substantially higher process temperatures than SnPb37 (left side of Figure 3.20). Lead-free SnAg3.5 solder is more expensive

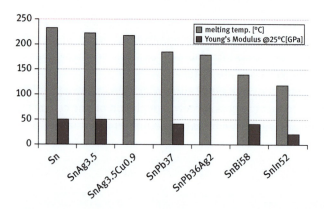

Fig. 3.20: Melting points of tin and common tin-based eutectic solders, Young's Moduli for tin and a selection of solders [44, 45].

than SnPb37 solder due to its increased material and process cost. The elevated melting point (221 °C) requires a substantially higher soldering temperature which in turn accelerates intermetallic diffusion and growth of intermetallic compounds. Since silver solubility in solid tin is below 0.1 % weight, it is mostly present as intermetallic compound Ag_3Sn.

Since the creep rate of SnAg3.5 solder is lower, higher stress levels will persist after solidification in the joint over a longer time period and reduce its strength. The Young's moduli for silver-based lead-free solders are substantially higher than for SnPb37, yet it is difficult to find consistent data in the literature. Low elastic moduli of solder alloys are favorable for cell stringing, since they reduce mechanical and thermomechanical stress in the joint and the solar cell. Elastic moduli increase with decreasing temperature, the alloys become stiffer.

Solders with low melting points have been proposed to combine the cell stringing process with the lamination process at temperatures around 150 °C or simply to reduce the thermomechanical stress associated with cell cooling from the solder's solidus point to room temperature. Two eutectic solder candidates are SnBi58 [43], with a melting point at 138 °C and SnIn52, melting at 118 °C. However, their low melting temperatures raise concerns with regard to their long-term reliability in PV modules.

3.1.6.2 Solder failure

Shear stress is the dominant mechanical stress acting on solder joints in solar cell strings. It originates from thermomechanical loads during string and laminate production and module operation. Purely mechanical shear stress arises from module deflection by static loads or wind.

Solder joints may fail due to fraction inside the solder material, although other types of fractions prevail in solar cell strings, e.g. at the solder/pad interface, inside the metallization, at the metallization/silicon interface or inside the silicon (Section 3.1.8.2). Starting points for fracture inside the solder are voids at solder-grain boundaries that grow by thermal diffusion and eventually merge to form microcracks. Cohesive solder fraction (in the bulk material) may arise from impact, sustained or cycling mechanical loads leading to instantaneous fracture, creep, fatigue, and combined effects [46]. Mechanical impacts may occur during string handling in module production, during module transportation, or mounting, due to harsh treatment. Also hail impact during operation has to be considered. If the joint is not substantially undersized, impact-induced fracture will usually occur outside of the bulk solder material.

Creep is related to plastic deformation (strain) of the solder material under persistent stress. For studying the reliability of solder joints, the homologous temperature (T_{hom}) is a usefull concept. It indicates the ratio of the material temperature with respect to its solidus temperature, both in Kelvin. Rising homologous temperature increases the disposition towards thermally activated processes like creep and grain growth. Solder used in PV module assembly are usually exposed to homologous temperatures above 0.5 already at ambient temperatures. At 85 °C, a common upper testing temperature for PV modules, SnPb37 solder approaches $T_{hom} = 0.8$, and SnBi58 even $T_{hom} = 0.9$. Creep behavior also strongly depends on the alloy composition and microstructure, while the latter may also change under mechanical stress.

SnCu0.5 behaves somewhat similar to SnPb37, if creep strength [MPa] and time to rupture is measured at 75 °C [47]. In contrast, less ductile, lead-free silver alloys SnAg3.5 and SnAg3.8Cu0.7 display substantially higher time to rupture ($100 \times ... 1000 \times$) and creep strength ($3 ... 8 \times$).

Isothermal fatigue is induced by cyclic mechanical loads. PV modules subjected to wind loads may respond with vibrations at frequencies of several tens of Hertz and thus induce cyclic mechanical loads on the joints.

Thermo-mechanical fatigue (TMF) is induced by cyclic thermal loads. Cyclic loads in PV module operation originate from temperature changes combined with CTE mismatch between joined materials as well as from intrinsic CTE mismatch between different phases inside the bulk solder. The cyclic load introduces shear and tensile stress in the material, which causes crack formation and growth. Especially during high temperature phases of the cycle, creep will contribute to TMF. When comparing TMF lives of SnPb and SnAg solders under defined strain, the lead-free solders tend to show an increased strength [48].

3.1.7 Electrically conductive adhesives

Electrically conductive adhesives (ECA) are being used in the electronics industry for interconnection and mounting in several applications where their advantages over soldering outweigh their higher cost. ECAs are often preferred in applications with high operating temperature or with elasticity requirements. They are lead free and do not require the use of flux, thus allowing clean processes.

ECAs consist of a polymer matrix and metal particles used as electrically conductive filler. Common ECA polymers include epoxies, acrylates, and silicones. The metal particles may be flake-shaped or spherical and consist of silver, silver coated copper, or tin-based alloys.

Isotropic electrically conductive adhesives (ICAs) are typically cured at temperatures between 120° and 180 °C without external pressure requirements. The contraction forces of the curing polymer matrix turn the material conductive. ICAs need high filler contents in the range of 70–95 wt % to overstep the percolation threshold [49], where the metal particles establish electric contact and conductivity sets in.

Since ICAs are processed at temperatures lower than common SnPb or SnAg solders, less thermomechanical stress is induced in the cooling phase by CTE mismatch, and interconnection of thermally sensitive cells, e.g. of cells based on heterojunction technology, is possible. Figure 3.21 shows volume resistivity ranges for different types of ICAs. Resistivities of solder alloys typically range within 12–14 µΩcm. While series

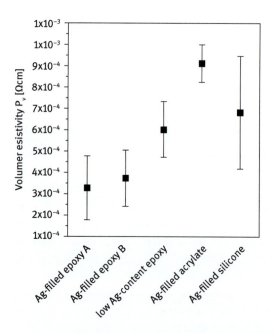

Fig. 3.21: Volume resistivity ranges for different ICAs [50].

resistance losses are relatively small in soldered contacts, these losses require special attention in glued contacts.

Anisotropic conductive adhesives (ACAs) only contain small amounts of filler materials, far below the percolation threshold. In turn they require considerable pressure to establish conductivity in the direction of the acting force.

In long-term operation at elevated temperatures in a humid environment, the conductivity of ECA joints can decrease due to oxide formation and galvanic corrosion. ECA pastes can be applied by screen printing, dispensing (Fig. 3.22) or jet printing on the contact pads of the solar cell or, in case of back contact cells, on a conductive backsheet.

Fig. 3.22: Solar cell with dispensed conductive adhesive line and ribbon (Fraunhofer ISE).

ECAs are often cured in two steps to improve compatibility with the module production process. Quick precuring of the ECA is thermally induced right after component placement, while final curing is conveniently achieved within the module lamination process which takes several minutes at temperatures of about 150 °C.

Challenging questions for the use of ECAs in module mass production are related to its higher material cost, to handling issues in module production and to long-term reliability in module operation.

3.1.8 Joint characterization

3.1.8.1 Peel test

The electric joints established by soldering or glueing need to achieve sufficient strength to withstand mechanical and thermomechanical loads during module production and the entire module service life. The first and very severe load coming into effect is the thermomechanical load that arises from CTE mismatch in the process of cooling after joint formation.

Joint strength is determined by the weakest link in the material stack from ribbon to wafer. This stack usually comprises copper ribbon, solder, porous silver, glass (Section 2.9), the bulk silicon and various intermediate layers.

A common procedure for assessing joint strength is the peel test. Figure 3.23 shows the peel test setup with the ribbon being peeled off from the solar cell. The cell can be fixed with a downholder or laminated on glass, as shown in Figure 3.23.

During the procedure, the force is registered together with the movement of the tool. In the peel process, the copper ribbon is strained. The displacement of the peeling tool has to be corrected for this strain to obtain the actual advancement of the peel front.

The peel test is performed at a constant angle and speed. Different angles lead to slightly different peel force results [51]. For reproducible results, a defined storage time after joint formation has to be observed. Due to stress relaxation, the joint strength continuously increases for some time after formation, at a particularly fast rate shortly after cooling.

If peel forces are translated to fracture energy, which is a geometry-independent parameter that describes the energy required to break the interfacial bondings at the peel front, more specific strength information can be gained [51].

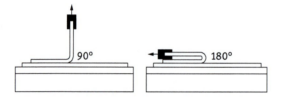

Fig. 3.23: Schematic peel test configurations for measuring peel force F at different peel angles.

Solar cells are supposed to provide a minimum peel force of 1 N per mm joint width for soldered contacts, as required by standard DIN EN 50461 [52].

Aside from the mere peel force, the fracture pattern provides important information about the joint. The identification of the fractured layer and the material analysis on the exposed surfaces helps to reveal the fracture mechanism. Common unaged soldered joints on screen printed metallization will usually display cohesive fracture within the porous silver layer. If the metallization is completely lifted off from the wafer, if silicon chips or parts of the wafer break out, this may indicate defects originating from cell or string production. In contrast, solder joints on plated metallization do not usually show cohesive fracture.

3.1.8.2 Metallography

Metallography is a method widely used in the electronics industry to study the composition and morphology within the layers of soldered joints. Joints on solar cells are difficult to prepare, since they involve brittle materials (silicon) as well as ductile materials (copper, tin). The preparation of a soldered cell sample (Figure 3.24) starts with cut-

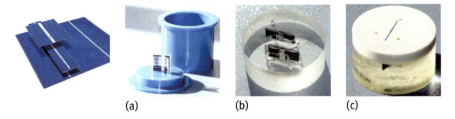

(a) (b) (c)

Fig. 3.24: Preparation steps for metallographic analysis of solder joints [38].

ting the solar cell alongside the busbar and then isolating the segment of interest (left side of Figure 3.24) by using a diamond wafer saw with continuous sample cooling.

The section is then embedded into a dual component resin and cured (Figure 3.24a,b). Figure 3.24c shows a soldered cell sample embedded in epoxy resin. In its upper part, the epoxy contains a ceramic powder as filler to increase hardness and to help achieve sharp sample edges in the following process. After curing, several grinding and polishing steps are required to reach the desired spot and to allow meaningful imaging by light microscopy, scanning electric microscopy (SEM), and energy-dispersive X-ray spectroscopy (EDX). If required, additional etchants are used to improve the contrast between different phases. SEM allows much higher magnification (> 10,000×) than optical microscopy (up to 1500×). With EDX, element maps of the cross sections can be obtained. They are valuable for identifying species and their diffusion. Both SEM and EDX require conductive embedding material.

Figure 3.25 shows a solder joint cross-section with the copper ribbon on top and the silicon wafer at the bottom. A thin dark Cu_3Sn-phase has grown immediately ad-

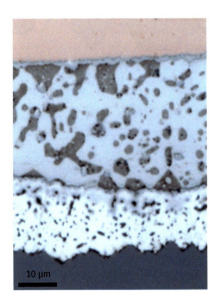

Fig. 3.25: Optical microscope image of solar cell joint with SnPb36Ag2 solder after heat storage at 130 °C for 85 hours (Fraunhofer ISE).

jacent to the copper, underneath follows a thicker and lighter Cu_6Sn_5-phase. The bulk SnPb36Ag2 solder contains lead-rich phases (dark) and tin-rich phases (light). The top layer of the cell metallization evolved into an Ag_3Sn-phase.

Figure 3.26 shows a SEM cross-sectional view of a solder joint with EDX element mapping. The copper ribbon is located on top, followed by a BiSn41Ag2 solder layer, the silver metallization, and the dark silicon wafer at the bottom. The solder itself shows two phases. The redish phase is rich in tin, as indicated by the EDX element map, and the greenish phase is rich in bismuth.

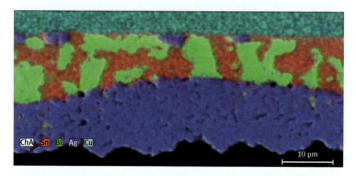

Fig. 3.26: EDX image with element mapping after metallographic preparation of a solar cell joint soldered with BiSn41Ag2 (Fraunhofer ISE).

3.2 Covers and encapsulants

3.2.1 Front cover

Wafer-based silicon solar cells are usually covered with a glass panel. Soda-lime glass (also called soda-lime-silica glass) has proven to be the most favorable solution for mass production, with its combination of high transmittance, mechanical strength, reliability and cost. For special applications that require extremely low weight or that are not exposed to outdoor conditions, transparent polymers have been proposed as substitutes, e.g. fluoropolymers, polymethyl methacrylate (PMMA), or UV-stabilized polycarbonate (PC).

The chemical composition of soda-lime glass comprises silicon dioxide (about 70 %), sodium oxide, calcium oxide and other additives. Glass panes produced in a float process on a tin bath (**float glass**) exhibit optically smooth surfaces, as known from architectural glazing applications. The two sides of a float glass pane display different bonding properties, which may be relevant for laminate or sealant adhesion. The side that was in contact with the molten tin can be identified under UV radiation, where it shows fluorescence.

For wafer-based PV modules, mostly **patterned glass** is used, also called "rolled glass". It is extruded between rollers, leading to slightly or deeply structured surfaces. Matt surfaces exhibit structures with maximum height differences of about 10 µm, while deeply structured surfaces may exhibit a 50 µm height difference and more. Patterned glass scatters the reflected light and thereby reduces its luminance, when compared to specularly reflecting float glass. This effect may reduce glare from PV modules in the field of vision. Only special patterns are able to reduce the reflectance of the glass surface; common matt glass surfaces have no respective effect.

Widely used architectural glass, recognizable by its greenish edge, contains about 500–1000 ppm iron which would generate effective absorption losses for PV applications of several percent. In solar applications, glass with very **low iron** content (about 100–150 ppm) is required. Iron can appear in glass as ferric (Fe^{3+}) and as ferrous iron (Fe^{2+}). Ferrous ions in the glass lead to a broad absorption band centered around 1050 nm, which can seriously reduce the performance of PV modules. Ferric ions absorb in the range of 385 nm with less detrimental effects on PV performance. To obtain solar grade glass which typically absorbs less than 1 % of the usable solar radiation, a low iron content is required and measures are taken to oxidize the residual iron from ferrous to ferric.

Since the thickness of the glass layer also influences absorptance, thinner glass is preferable for module efficiency reasons. In glass-backseet module designs, the glass pane is usually 3–4 mm thick. If glass is also used as rear cover, both panes are typically 2 mm thick. For light-weight PV modules, thin glass is available down to 0.85 mm thickness.

A second very important property of the glass panel is its mechanical strength. Being a brittle material, the strength of the glass is limited by surface flaws, and tensile stress is much more critical than compressive stress. Statistical approaches (Weibull statistics) are used to characterize flaw population and fracture strength [53].

Annealed glass, the common architectural glass quality manufactured through slow cooling, displays low levels of surface compression and low characteristic bending tensile strength in the range of 45 N/mm². By controlled quick cooling with a starting point above 600 °C, glass can be **thermally strengthened**. In this process, the glass surface is compressed due to the delayed cooling and associated contraction of the inner layer. The compression results in an increased bending tensile strength. Its mechanical strength above 70 MPa reduces breakage risk under hail impact or heavy mechanical loads. For wafer-based PV modules, **thermally toughened glass** is commonly used. It provides a strength of 120 MPa due to its even higher residual stress, but it shows poorer planarity. Spontaneous breakage risks caused by NiS inclusions need to be ruled out.

Due to the inherent mechanical stress, strengthened glass cannot be cut or drilled once it has been strengthened – the entire pane would break. This means that glass panes need to be delivered to the module production line in their final formats. When broken, thermally strengthened glass displays the same breakage pattern as annealed glass, comprising large pieces with sharp edges. In contrast, a toughened glass pane shatters completely into small granules.

Strengthened and toughened glass is also more tolerant against temperature gradients across the glass plane than annealed glass. Critical gradients may occur in PV module operation e.g. under partial shading or hot-spot conditions (Section 5.2).

Only recently, thermal strengthening processes have been adapted to a glass thickness below 2 mm, where it can achieve a typical bending tensile strength of 120 N/mm^2. Until a few years ago, glass thinner than 2 mm had to be chemically strengthened by exchanging surface sodium ions with potassium ions in a bath of molten salt. This process is much more expensive than thermal treatment.

The glass surface also deserves special attention, since it causes optical losses through reflection. The reflectance of an interface between two materials (e.g. air/glass) is determined by the square of the difference of their refractive indeces (equation (5.15) in Section 5.5.1). Therefore it is advantageous to introduce one or more additional layers with intermediate refractive indeces, called **antireflective coatings** (ARC). Although one or more reflecting interface are newly created, the sum of the interface reflectances will be lower than the initial single interface reflectance. The intermediate layers or coatings provide a step-wise reduction of index discontinuity. Figure 3.27A shows the initial situation with a relatively large index discontinuity leading to a reflectance of 4.3 %. There are three important AR concepts which come into question at the air/glass interface of PV modules as well as on the encapsulant/silicon interface.

The first concept introduces one or more so-called "thick layers", as shown in Figure 3.27B and C. The supplement "thick'" indicates that the layer thickness exceeds half of the coherence length of sunlight. In consequence, no coherent interaction is taking place between light reflected from different interfaces and the reflected intensities from each interface add up in terms of power. When silicon wafers are encapsulated in a polymer, the entire encapsulant can be regarded as a thick AR layer with intermediate, though not optimal refractive index. Another practical example is a module laminate with a fluoropolymeric cover (n = 1.4) separating ambient air (n = 1) and the encapsulant (n = 1.48). In practice, it is difficult to provide coatings with refractive indeces much smaller than 1.4, approaching the ideal value of 1.23. One solution to this challenge is a so-called **effective optical medium** which provides air-filled subwavelength structures (e.g. pores) within a higher index matrix material. The pore density controls the effective index of refraction.

In theory, an increasing number of intermediate layers can be introduced such that the refractive index ratio (n_{i+1}/n_i) for adjacent layers is always equal. Ultimately this iteration approaches the second concept: a **gradient index** layer where the index of refraction changes continuously between two materials and reflectance is reduced to zero (as shown in Figure 3.27D). In practice, this concept can be approached by a gradient in the subwavelength pore density in an effective optical medium which will result in a gradually increasing mean material density and a corresponding refractive index gradient. Examples for such structures are the needles etched in silicon as

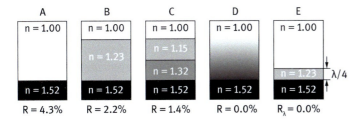

Fig. 3.27: Different idealized AR concepts for an air/glass interface with corresponding total reflectance R: initial air/glass interface (A), one thick AR layer with intermediate refractive index (B), two thick AR layers with intermediate refractive indeces (C), effective medium with gradient index layer (D) and $\lambda/4$ layer (E).

shown on the right side of Figure 2.22 and AR sol-gel layers on module glass. Both AR layer concepts mentioned, the thick layers and the thick gradient index layer, work independently of wavelength and path lengths.

The third concept, quarter-wave AR layers ($\lambda/4$), have to be tailored for a reference wavelength and ray angle within the layer. An ideal $\lambda/4$ AR layer has an intermediate refractive index such that the refractive index ratio is equal at both interfaces. The reflected waves then have the same amplitude but a phase shift of $\lambda/2$ which causes destructive interference. The resulting reflectance is zero for the reference wavelength (Figure 3.27E). At different wavelengths and incidence angles, reflectance increases.

Figure 3.28 shows SEM-images of two nanoporous $\lambda/4$ AR layers. Nanoporous AR layers on glass are most commonly applied in a sol-gel process, some manufacturers use magnetron sputtering or glass etching. The main material constituting the layer matrix is SiO_x. The most effective antireflective coatings used in PV applications can reduce the orthogonal air/glass interface reflectance to about 1 %. Today, most of the PV cover glass is equipped with an ARC.

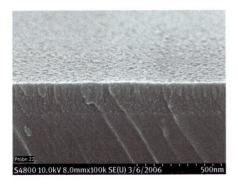 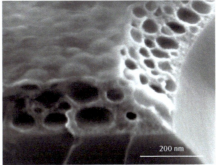

Fig. 3.28: SEM cross-sectional view of nanoporous AR layers displaying fine pores (left image, by SGS INSTITUT FRESENIUS GmbH, AR coating on glass, reprinted with the permission of fsolar) and large, closed pores (right image, AR coating sampled on silicon, reprinted with the permission of DSM).

The reflectance of a flat air/glass interface also depends on the angle of incidence (Section 5.5.1). At oblique incidence, the reflectance for unpolarized light increases. Figure 3.29 shows reflectance curves of an air/glas interface coated with 0 to 3 ideal AR "thick" layers, where the refractive indeces of subsequent layers always increase by a constant factor. This effect is relevant for the electric yield of stationary mounted PV modules and in climates with large fractions of diffuse irradiance.

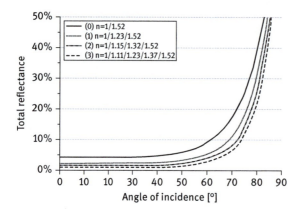

Fig. 3.29: Total interface reflectance of an air/glass interface for unpolarized, monochromatic light and 0 to 3 intermediate layers; simple model neglecting higher order reflections and coherence effects.

A different approach for reducing reflectance is taken by inverted pyramid structures (Figure 3.30) on the glass surface [54] of patterned glass. They increase the number of ray intercepts on the glass surface by generating multiple reflections.

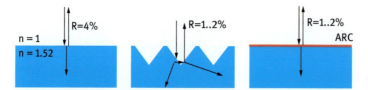

Fig. 3.30: Schematic drawing showing reflection from flat glass surface (left), inverted pyramid structures (middle) and coated glass surface.

The AR treatments discussed are also effective at oblique incidence at and beyond 60° and may increase the annual module yield by several percent [55].

Depending on the local aerosol deposition rate and quality, the frequency of rain fall and the module tilt angle, soiling can severely reduce the transmittance of the module front cover and the electric yield. Different **antisoiling coatings** on glass are offered with the intention of creating superhydrophilic surfaces that are easily wetted and rinsed by water at contact angles below 10° or super-hydrophobic surfaces that impede the adhesion of water at contact angles above 120° [56].

While most glass coatings require elevated temperatures for curing, some products may be applied at ambient temperatures and may thus be used even for module retrofitting.

3.2.2 Rear cover

Most manufacturers of wafer based c-Si modules use polymer backsheets as rear cover, which typically consist of 2 or 3 single sheets laminated with polyurethane (PU) interlayers (Figure 3.31). For several years, the market had been dominated by 3-layered backsheets comprising one polyethylene terephthalate (PET) film sandwiched between two polyvinyl fluoride (PVF) films. The PET film of typically 150–300 microns thickness provides **dielectric breakdown** protection, while the thin PVF films of several tens of microns thickness improve durability.

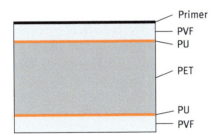

Fig. 3.31: Schematic cross-section of a backsheet, total thickness 350 µm; the primer improves adhesion to the encapsulant.

Today, a wide range of backsheet types are offered, including polyvinylidene fluoride (PVDF), special grades of PET or polyamide (PA), and several types do not use PVF any more or totally forgo fluoropolymers.

Backsheets are responsible for electric isolation between the cell string and the rear environment of the module. Depending on the system voltage, the potential of the cell string versus ground may reach 1000–1500 V. The thickness of the PET film is chosen to provide the appropriate protection.

A second very important backsheet property is the temperature-dependent **water vapor transmission rate** (WVTR). The equilibrium water intake of the module depends on the WVTR and the ambient conditions. Low water intake is favorable for a long module service life, since it slows down degradation by hydrolysis and oxidation. During daytime, solar irradiance causes the module to heat up, and the increased

WVTR will facilitate water release from the module into the ambient air. After sunset, temperature falls, and ambient humidity usually rises, but the reduced WVTR will counteract water intake.

In addition to this daily fluctuation, in many locations modules also encounter seasonal fluctuations. For PV cell technologies that demand a particularly high degree of humidity protection, backsheets can be equipped with an additional high-barrier interlayer made of silicon oxide or aluminum.

Backsheets not only impact electric safety and service life, they also influence module efficiency and performance. White backsheets are advantageous in two respects. First, they partially recycle light that falls nearby the solar cells by scattering it back to the glass-air interface from where it may be reflected onto the active cell surface. This effect may lead to optical gains of 1–3 % (Section 5.5.7).

Secondly, a white module rear side absorbs less light from the environment, which helps to reduce module operation temperature. For both effects, a high reflectance on both sides of the backsheet is advantageous. White backsheets can achieve up to 85–90 % effective reflectance.

Common wafer-based modules exhibit substantial inactive areas inbetween the cells, around the module borders and in the string connection regions. By introducing strips of structured, specularly reflecting film that cover the uniformly scattering backsheet surface, higher fractions of light can be reused (Figure 3.32). These structured reflectors may show a similar cross-section as the sawtooth profiled ribbon in Figure 3.6. The strips may also be used to cover a flat ribbon, thus achieving a similar functionality as a structured ribbon (top left in Figure 3.6). A subsequent dark appearance of inactive ares with mirror images of the active cell surface is an indication of high optical module efficiency.

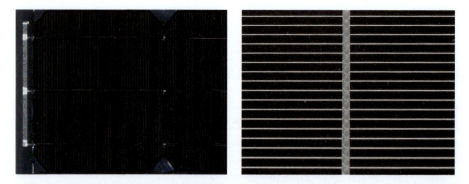

Fig. 3.32: Module sample with monocrystalline, pseudo-square solar cells, where inactive areas on the ribbon and around the cells have been covered with structured specular reflecting film (left); detail of film covered ribbon (right) showing mirror images of the adjacent metallization fingers (Fraunhofer ISE).

Some manufacturers use glass as the rear cover instead of polymer backsheets. Two glass covers provide tight diffusions barriers over the module area. If they have similar thickness, the cell matrix is located in the neutral plane of the compound (Figure 3.33), thus experiencing less tensile stress when the module is deflected by external loads. In a glass-foil module, the cell matrix is beyond the neutral plane and is subjected to tensile stress when the module bends under front side pressure. In the past, thermally toughened module glass was only available with a thickness at or above 3 mm. With two times 3-mm-glass, module weight may become a concern. Now that thinner glass is available, glass-glass modules are becoming more popular. A PV module with a glass front and rear cover with a thickness of 2 mm has several advantages over a glass-foil module with a 4 mm single sheet, without increasing the module weight.

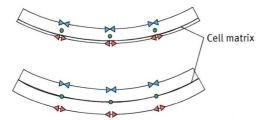

Fig. 3.33: Schematic cross-section of a bended glass-foil-laminate (top) and glass-glass-laminate (bottom), showing the maximum compression fiber (marked blue), the neutral fiber (green), and the maximum elongation fiber (red).

Different approaches are taken in glass-glass modules for connecting the cables to the cell matrix. If common junction boxes are to be used, the rear glass requires holes. Yet, glass drilling is expensive and reduces the mechanical strength of the glass. Alternatively, a cut-out in the rear glass or a projecting front glass can provide means for the box fixation. A totally different approach uses edge contacts that do not require any measures on the glass side. In this case, the bypass diodes have to be placed inside the laminate (Figure 3.38).

Glass-glass modules are used in building integrated PV (BIPV) applications, where they can provide safety glass properties. Some PV module manufacturers have received approvals that their products based on PVB or EVA encapsulants (Section 3.2.3) comply with respective building codes. Bifacial modules require glass or transparent polymer films for rear cover.

3.2.3 Encapsulants

Polymer encapsulants are used as intermediate layers between the cell matrix and the module front and rear cover. The encapsulant material strongly adheres to all surfaces, binding the components mechanically into a laminate. High volume resistivity is required to inhibit leakage currents. Transmission rates for water and oxygen need to be sufficiently low to inhibit corrosion. The front side encapsulant layer has to provide nearly 100 % transmittance in the relevant wavelength range. During a service life of about 25 years, encapsulants have to resist substantial doses of UV irradiation combined with elevated operating temperatures and a certain level of humidity inside the module.

Very importantly, the encapsulant needs to compensate shear stress between all components that arise due to temperature changes in conjunction with CTE mismatch (thermos-mechanical loads, Section 2.8) as well as due to mere mechanical loads perpendicular to the module plane (Figure 3.33). If module temperature is changed in the interval of −40 °C to +85 °C within the IEC test [57], and we assume the cell center as a fixed point, strainless length difference between a 156 mm solar cell and the corresponding glass cover can exceed 35 µm, depending on the distance to the cell center (Figure 3.34, left). Another simplified calculation shows that if a module of 1600 mm edge length is deflected by 100 mm relative to its short axis and the cell matrix is 1.5 mm behind the neutral fiber, the displacement can also exceed 35 µm.

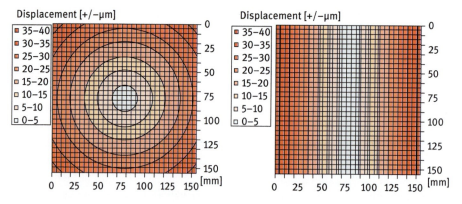

Fig. 3.34: Calculated strainless displacement between a silicon solar and soda-lime-glass for temperatures in the range of −40…+85 °C (left) and for a module deflection of 100 mm with respect to a bending axis parallel to the short module edge (right).

From these considerations it follows that the encapsulant should display a sufficient layer thickness and a moderate shear modulus G (and consequently a moderate elastic modulus E); otherwise the solar cells may experience critical tensile stress under mechanical loads from the front side [58].

Special attention is required if the glass transition temperature [59] of the material lies within the operation temperature range of the module, since the transition causes severe moduli changes (Figure 3.35). The dependence of encapsulant moduli on temperature is measured by dynamic mechanical analysis (DMA). Different measurement frequencies usually lead to slightly different results in T_G. In module operation, not only quasi-static shear stress occurs due to CTE-missmatch, but wind induced module vibrations in the range of several tens of Hertz impose dynamic stress.

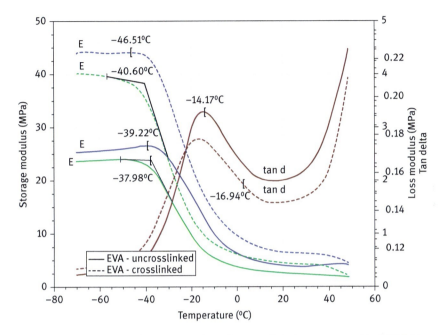

Fig. 3.35: Moduli for initial (thermoplastic) and for cross-linked EVA measured by DMA [60].

The encapsulant material is expected to establish strong bonds to the neighboring materials. The adhesion strength achieved is checked in another off-line and destructive test, the peel test (Section 3.2.5.1).

The front side encapsulant layer needs to be highly transparent in the relevant spectral range between 380 and 1100 nm; typically losses at or below 1–2 % are expected. This specification also restricts the choice of polymer additives, especially of UV absorbers.

High-volume resistivity is required to inhibit leakage currents between the cell matrix and grounded parts where voltage differences of up to 1000–1500 V may occur. As far as the module frame and edge is concerned, leakage currents are a safety issue and addressed by the wet leakage current test within IEC 61215 [57]. In conjunction with the front glass, leakage currents may arise due to wetting by dew or rain, which sets the surface to ground potential. In the resulting electric field between the outer grounded surface and the cell matrix, mobile ions can support a leakage current from the glass over the encapsulant into the cell and cause severe temporary or permanent cell efficiency losses (**potential induced degradation**, PID).

When modules are assembled, the encapsulant is required to flow and completely fill all gaps of the cell matrix. The viscous liquid state is usually achieved by melting a thermoplastic polymer originally provided as a sheet. Later in the process, the thermoplastic encapsulant may be cross-linked into an elastomer, which prevents any flow at elevated temperatures. Especially silicone encapsulants may also be provided initially in a liquid state; in this case, cross-linking is performed after the material has embedded the cell matrix. Depending on the choice of materials, primers may be required to achieve strong bonds between encapsulant and cover materials.

Initially, silicone (polydimethylsiloxane, PDMS) and polyvinyl butyral (PVB) were used to encapsulate commercial PV modules. Silicone application was limited for reasons of cost, while PVB induces reliability risks outside of a hermetically sealed laminate. In the late 1970s, the copolymer ethylene vinyl acetate (EVA) was investigated and developed as a new PV pottant [61]. It soon became the prevalent material used for wafer-based module encapsulation, and has remained so ever since. Provided initially as a thermoplastic foil with no insoluble fraction, it is first melted and then cross-linked to an elastomer in the lamination process. A minimum gel content specified by the material manufacturer has to be achieved during lamination (Section 3.2.5.2); typical limits are defined around 70 %–80 %. EVA encapsulants contain several additives [62]. Silanes, usually alkoxysilanes, promote the adhesion to glass with their silanol groups. Peroxides initiate the cross-linking process at temperatures above 140 °C. Additionally, they initiate the stable adhesion of silane to the glass surface by covalent Si-O-Si bonds. Different stabilizers absorb UV irradiation or serve as antioxidants [63].

Prior to the introduction of EVA, polyvinyl butyral (PVB) was often used as module encapsulant. PVB is synthesized from polyvinyl alcohol by reaction with butyraldehyde. The bonding to glass occurs through the polar alcohol groups via hydrogen bonds, and plasticizers are added to reduce Young's modulus. PVB is widely used in architectural and automotive glazing to manufacture safety glass. In case of glass breakage, a PVB layer in between two glass panes can sustain the broken glass for a certain time period even under load. Building integrated PV (BIPV) may require this property to reduce the injury risk from broken glass fragments. PVB lamination is usually performed in an autoclave process without any cross-linking, thus preserving thermoplastic properties of the encapsulant.

Ionomer encapsulants provide lower water vapor transmission and absorption rates than EVA or PVB combined with a high elastic modulus. They are used in glass-glass modules to improve humidity protection. In a symmetric glass-glass build-up, the cell matrix is located in the neutral plane and even a high modulus encapsulant imposes less risk for critical cell stress. A thermoplastic polyolefin elastomer (TPO) consists of a thermoplastic polyolefin (e.g. polypropylene) with some elastomer fraction (e.g. EPDM).

Silicone encapsulants based on polydimethylsiloxane had been used in the early days of PV module technology. They provide an excellent intrinsic durability against UV and thermal induced degradation, coupled with favorable mechanical properties as low glass transition temperature and Young's modulus. The high Si-O bond dissociation energy cannot be delivered by photons from terrestrial UV irradiation. Yet, silicones did not make it into mass production due to their higher cost.

Figure 3.36 displays the chemical structure of different encapsulant materials. Table 3.1 gives an overview on material types available for module encapsulation.

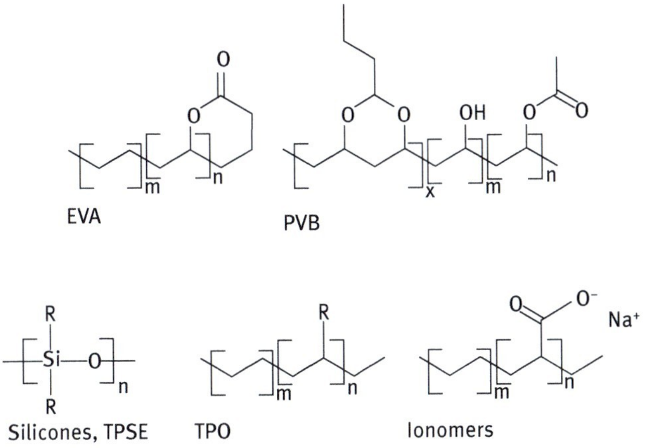

Fig. 3.36: Chemical structure of materials used or tested for module encapsulation.

Tab. 3.1: Encapsulant materials overview [64].

	EVA	PVB	Ionomer	TPSE	TPO	Silicone
material type	ethylene-vinylacetate copolymer	polyvinyl butyral	ionomer	thermoplastic silicone elastomer	thermoplastic polyolefin elastomer	1K/2K silicone
delivery	sheet	sheet	sheet	sheet	sheet	liquid
final state	elastomer	thermoplast	thermoplast	thermoplastic elastomer	thermoplastic elastomer	elastomer
refractive index	1.48	1.48	1.49	1.42	1.48	1.4
glass transition temp. [°C]	−40...−30	12...20	40...50	−100	−60...−40	−50
Young's modulus [MPa]	< 68	< 11	< 300	< 280	< 32	< 10
Volume resistivity, dry [Ohm.cm]	1E14...1E16	1E10...1E12	1E16	1E16	1E14...1E18	1E14...1E15

3.2.4 Edge sealed designs without encapsulant

In order to save the process time and material required for embedding the cell matrix in encapsulant layers, alternative module designs have been proposed. They use a dual edge seal to join a front and a rear glass pane. Very similar to architectural insulating glazing, the primary polyisobutylene seal is responsible for inhibiting moisture diffusion, while the secondary seal, usually a silicone, strengthens the glass joint mechanically. Since the solar cells are not optically coupled to the front glass, these designs require double side AR coated glass and cells with particularly low front side reflectance.

The edge sealed module designs follow different approaches for cell matrix fixation. The "NICE" concept [65] makes use of the atmospheric pressure not only to keep the cells in place, but also to maintain cell interconnection. The pressure inside the module is a few hundred mbar lower than atmospheric pressure. The resulting pressure difference maintains mechanical and electrical contact between interconnection ribbons and cells, and therefore no soldering is required.

The "TPedge" approach (Figs. 3.37, 3.38) uses tiny dots of elastomeric material to secure the cell matrix to the rear glass. On the front sides, pins of the same material are applied to space the cells relative to the front glass. Mechanical loads are transferred through pairs of corresponding pins from the front to the rear glass. The cell matrix is assembled in a common manner by soldering. No pressure difference between the

Fig. 3.37: Schematic cross section (left), top view (middle), and bottom view (right) of a TPedge module design.

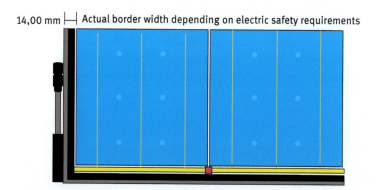

Fig. 3.38: Schematic top view of a TPedge design with module corner showing integrated bypass diode (bottom center) and module cable (left).

interior and exterior is required. Compared to the material volume for full encapsulation, the pins only require a material volume of about a thousandth. The sealing process, which takes less than one minute, is substantially faster than lamination.

3.2.5 Laminate characterization

3.2.5.1 Peel test

Peel tests are used for quality assurance and degradation assessment of laminate material interfaces in PV modules. Procedures are described in ASTM D6862-11 Standard Test Method for 90 Degree Peel Resistance of Adhesives [66] and in DIN EN 28510-1 Adhesives – Peel test for a flexible-bonded-to-rigid test specimen assembly [67]. The module front glass provides the rigid support from which the polymer layers (backsheet, encapsulant layer, or both) is peeled off. For the peel test preparation, 2 parallel cuts of 200 mm length at 10 mm distance are conducted with a knife or a laser tool. The location and the depth of the cut as well as the starting point preparation determine the interface under test, which may be the backsheet adhesion to the encapsulant or the encapsulant adhesion to the cell or to the front glass. The head of the strip has to be detached and clamped in the test apparatus (Figure 3.39). In order to easily detach the backsheet from the encapsulant, a nonlaminated starting point for the peel has to be provided by placing separating interlayers between the sheets before lamination.

During the peel process, a constant peel angle of usually 90° and a constant peel speed are maintained while the peel force is registered. Due to elastic and plastic deformation of the peel strip under tensile stress, the peel progress on the laminate is slower than the displacement of the clamp.

While the peel test primarily addresses the bonds between the encapsulant and its adjacent materials, it may lead to adhesive or cohesive breakage of other components, e.g. a breakage inside the rear side cell metallization, silicon wafer chipping, a separation of backsheet layers, or a rupture inside the encapsulant.

The strong adhesion that can be achieved between EVA and glass in excess of 10 N per mm strip width requires careful preparation and interpretation of the peel tests [68]. In the course of the peeling process from glass, the force will usually increase and reach a maximum value within the first 1–2 cm; this value is than reported as the adhesion strength. If EVA is peeled without any attached backsheet from glass, the EVA strip can be excessively strained and the measurement will not deliver meaningful results for the adhesion strength. The results will also be distorted if the backsheet adhesion to EVA or the internal adhesion in between backsheet layers is weak. If the encapsulant adhesion to the outer backsheet layer is stronger than the internal adhesion inside the backsheet, the backsheet delaminates during the test and it is not possible to measure the adhesion between backsheet and encapsulant.

In most cases, the encapsulant adhesion to the cell rear side is stronger than the cohesive strength of the rear-side cell layers, e.g. of screen-printed aluminum. In consequence, rear-side cell layers will break. Locally intact metallization after peeling is an indication of gas inclusion. On the solar cell rear side, the peel test therefore gives orientative information on the strength of the cell metallization layers and not on encapsulant adhesion. When the peel crosses a cell gap, the peel force sharply increases. In these gaps, the peel force tears the encapsulant, possibly separating the initially distinct encapsulant sheets, and may peel the encapsulant from the glass, depending on the weakest point. The maximum peel force in the cell gap is reported (left side of Figure 3.39). Only at module borders, where no cells are present, or in dedicated laminate samples without cells, the peel test precisely address the encapsulant-glass adhesion.

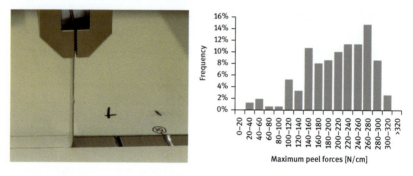

Fig. 3.39: Peel test in progress on a module crossing a cell gap that appears dark and a shiny ribbon (left) and histogram of measured peel forces (right) on a large number of commercial modules (Fraunhofer ISE).

3.2.5.2 Gel content

During lamination, EVA encapsulants are required to reach a gel content specified by the manufacturer. ASTM D2765-11 defines the gel content as the insoluble percentage by mass of a polymer for specified solvent and extraction conditions [69]. This insoluble fraction is determined offline in a destructive procedure which takes several hours. The cited standard describes the methods for determining the gel content and swell ratio of crosslinked ethylene plastics. For a precise determination it suggests the use of a Soxhlet extraction procedure.

Material samples for production quality assurance are taken from designated laminate samples that have been processed parallel to the module production. If such samples are not available, the module backsheet has to be opened to take samples, which destroys the module. To account for laminator inhomogeneity, samples are usually taken from different parts of the module, including central and peripheral regions.

The EVA sample is cut into small pieces, dessicated, weighted, and placed inside a porous filter thimble. The thimble is then introduced into the upper reservoir of a Soxhlet apparatus (Figure 3.40). A strong solvent (usually xylene [69] or toluene) in

Fig. 3.40: Soxhlett extraction apparatus with condensor helix on top, white filter thimble in the middle and heatable solvent reservoir at the bottom.

a lower reservoir is continuously heated to maintain it at the boiling point. Solvent vapor ascends, bypasses the upper reservoir, and reaches a condenser. From there, liquid solvent trickles into the lower reservoir and extracts soluble materials from the sample. Before cross-linking, 100 % of an EVA sample can be dissolved. After full cross-linking, only about 10 % (weight) are soluble, and the rest constitutes a gel.

As soon as the solution reaches the level of a syphon, the upper reservoir voids into the lower reservoir where the concentration of solubles increases over time. The extraction process is usually stopped after a minimum number of syphon spillovers.

Finally the remaining material inside the filter thimble is removed, dessicated and weighted again. The remaining weight fraction provides the gel content. The gel content is not to be confused with the degree of cross-linking, which denotes the molar fraction of cross-linked monomers.

Several alternative tests have been proposed to estimate the gel content and to provide quicker, more reproducible results [70]. Differential scanning calorimetry (DSC) identifies threshold temperatures for thermally active transitions including chemical reactions or phase changes. DSC results indicate residual initiator that has not been consumed during lamination. The solvent swelling method determines the solvent uptake of the polymer sample which depends in turn on the gel content. Dynamic mechanical analysis (DMA) reveals viscoelastic properties of the polymer. As nondestructive methods with inline testing capability, indentation through the backsheet and spectroscopic analysis [71] via the front glass have been proposed.

3.3 Junctions and frame

The junction box (Fig. 3.14) hosts the electric connection between the cell matrix and a pair of external cables. The cables usually provide a total conductive cross section of 4 mm² in order to achieve an electric resistance below 5 mΩ/m. They are made of tinned multistranded copper wires with a double layer jacket. The cables are equipped with standardized plugs with a specified contact resistance of typically below 0.5 mΩ.

The casing material of module junction boxes, e.g. a phenyl ether polymer (PPE), has to fulfil specifications related to mechanical stability, elevated temperatures generate inside, humidity, UV irradiation and flammability. Some boxes provide a pressure valve with integrated desiccant to reduce the humidity inside. Other boxes are filled with a sealing material after mounting on the laminate.

Fig. 3.41: Junction box PV-JB/WL-H from Multi-Contact (left) and Kostal Samko 100 01 (right) with tin-plated copper circuit, four connection points, three bypass diodes, and two connection cables. Reprinted with the permission of Multi-Contact and Kostal.

The box also contains diodes that bypass a cell substring when it fails to deliver a current of the same magnitude as other substrings. This failure may be due to full or partial shading of at least one cell, or to a malfunctioning cell. Figure 3.42 shows a situation with partial shading of one cell out of the 60 cells in the module. Without any protective measures, this cell could experience a reverse voltage in the range of the sum of all other cell voltages in the module. This sum would likely exceed the breakdown voltage of the weakly performing cell, which typically is in the range of −15 V. A hot-spot would be the consequence, which may lead to permanent damage

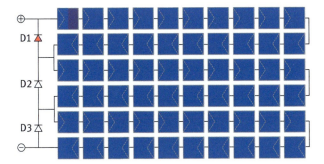

Fig. 3.42: Schematic drawing of a common 60-cell-module interconnection scheme with three bypass diodes on the left side and a partially shaded solar cell in the top left corner that activates the upper diode.

in the cell. The generated heat may additionally damage the module encapsulant and backsheet and even start a fire. Even if no material damage occurs, the weak cell would negatively affect the current in the entire serial circuit and cause severe yield losses.

A bypass diode limits the reverse voltage acting on the weak cell and may prevent permanent damage. In its active state (denoted red in Figure 3.42), the diode dissipates power according to the product of the current and the diode's voltage drop. In PV modules, Schottky diodes are used due to their small voltage drop in forward bias (equal to threshold voltage; see Chapter 2) of about 0.4 V. On the other hand, Schottky diodes dissipate more power under reverse bias, i.e. in regular module operation mode, than silicon diodes. The thermal design of the junction box needs to allow sufficient heat dissipation from the diodes to the ambient.

Figure 3.43 shows a picture taken in a PV power plant with a superimposed infrared image. Several cells in the left string of the marked module show considerably increased temperatures, while the central and right strings seem to be less or little

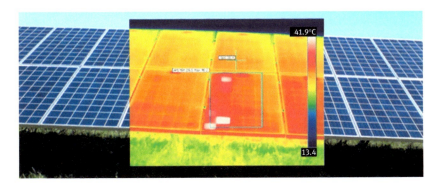

Fig. 3.43: Thermographic investigation of a solar generator in a photovoltaic power plant (Fraunhofer ISE).

affected. The warm spot close to the upper module edge is caused by the rear side junction box, and gives an indication that one diode may be dissipating power. The other modules also display an elevated temperature at the junction box, but this is mainly due to the rear side thermal insulation caused by the box.

The joints inside the junction box between the module's string connectors and the external cables deserve special attention. In a PV system that operates at 1000 V DC and 8–9 A, fretting corrosion or a loose connection can cause a stable electric arc that may dissipate large amounts of power. If the inverter does not detect arcing and subsequently interrupts the DC circuit, there is a severe risk of fire.

Modules may be equipped with active devices. Microinverters deliver AC voltage, which allows parallel connection directly to the grid. Since every module is driven at its individual MPP, mismatch losses are avoided. So called "power optimizers" also operate each module at its individual MPP, while delivering a DC current that is adjusted to the entire module string. This independent MPP operation reduces mismatch losses due to different performance, shading, or soiling levels within a string of modules. Different strategies are used to adapt the operating point. Some devices convert the module DC voltage to AC, transform it to a different voltage level, and then convert back to DC. This multiple transformation introduces some power loss even at times when no adaptation is required. Other devices analyze the IV curve of the module at intervals and only adapt the module's internal resistance when mismatch is detected. Active devices may also provide data for monitoring on individual module level.

So-called "active bypass diodes" are switching circuits based on field effect transistors (FET) that can be used pin-compatible to diodes. Their voltage drop in forward bias is lower by a factor of about 10, when compared to Schottky diodes. In consequence, they dissipate much less power when actively bypassing a low current string. This reduction not only facilitates thermal management, but also reduces the power loss in the system. When active diodes are operating in reverse bias, their leakage current is several orders of magnitude lower than for Schottky diodes (Figure 3.44). Since all diodes display a positive temperature coefficient in their leakage current, a low leakage current is favorable to prevent thermal runaway.

Due to the very low heat dissipation of active bypass diodes, they can be provided in a flat casing for integration into the module laminate. Such a module design does not require a central junction box. The connection cables can be attached to opposite module edges, similar to the design in Figure 3.38.

The integrated circuit of an active bypass diode may offer additional functionalities as lightning protection or a controllable module shut down by a short circuit. In case of maintenance or fire, this feature substantially improves the electric safety of the system.

The module frame has several functions related to stability, edge protection, handling, and mounting. It can be regarded as a reinforcement of the module against mechanical loads perpendicular to the module plane, originating from snow or wind. The frame encloses the sensitive edge of the module laminate and avoids direct contact

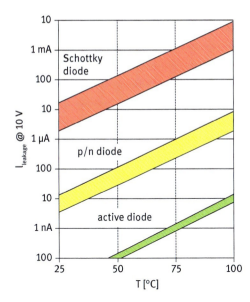

Fig. 3.44: Temperature dependent leakage current in reverse bias operation for passive and high efficiency active bypass. Image from Heribert Schmidt, personal communication.

with water. It also protects the fragile glass edge from mechanical impacts. Finally, the frame provides a convenient means for handling and mounting the module.

Most manufacturers use extruded aluminum profiles as frames (Figure 3.45) with a profile height of 33 to 50 mm, and apply tape or a sealant inside the notch to fix the frame to the glass edge.

If edge protection is less critical, modules can be mounted without frames. Installers may use pairs of clamps that attach to the edges or back-rail profiles glued to the rear side of the module. Back rails cannot be used with bifacial modules because they partially shade the rear side.

Fig. 3.45: Schematic cross-sections of an aluminum frame (left), a clamp (middle) and a backrail (right).

3.4 Module production

3.4.1 Production process

Module production can be configured as a fully automated in-line process. In the following a typical process chain (Figure 3.46) for glass/foil modules will be described.

At the beginning of the module assembly line, the glass panes are cleaned in a glass washer using warm deionized (DI) water and brushes. Freshly cleaned glass surfaces provide the best conditions for achieving strong adhesion in the lamination process and a high level of quality assurance (QA). The glass is dried in an air stream. Glass dimensions and the integrity of the glass edges are checked by a vision system. In case that the glass has been handled vertically up to this point, it is now turned and placed horizontally on a conveyor. The front glass serves as a support for the following module production. From the glass cleaning to the lamination, the production requires a low-dust environment.

The front side encapsulant sheet is cut to size and placed on the glass. For proper handling, the encapsulant surface is not allowed to stick to the glass. Storage conditions and shelf life for encapsulants in closed and open packages have to be carefully observed. Especially in case of EVA with its volatile additives, the material may have to be used within a few hours after unpacking, as specified by the manufacturer.

Parallel to the glass provision, stringers prepare cell strings of 10 or 12 serially interconnected solar cells. In the layup station, the strings are placed face-side-down on the encapsulant and interconnected, forming the cell matrix. The rear side encapsulant and the back-sheet are cut to size and placed on the matrix. Before lamination, the module connections are fed through the rear side sheets. This layup is then conveyed to the laminator and subsequently to a cooling press. After reaching room temperature again, the module edges are trimmed by removing salient encapsulant and backsheet. Then the frame tape is applied, which is intended to fix and seal the frame profiles to the glass edge. After frame profiles are mounted, the junction box is fixed with silicon

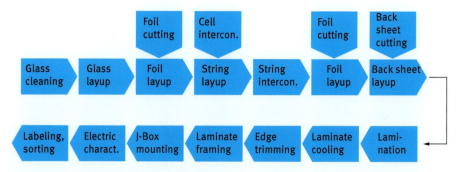

Fig. 3.46: Main process steps in module production.

or adhesive tape and connected to the module's projecting string connectors by either welding, soldering or clamping.

Some manufacturers fill the junction box with a sealing material, e.g. silicone. After this, the module is characterized at standard testing conditions in an inline flasher. The module data is printed on a label which is glued to the module rear side. As a last step, the modules are classified into bins according to their STC power.

3.4.2 Production equipment

Figure 3.47 shows a semiautomated module production line with a nominal production capacity of 70 MW/a operating counter-clockwise. At the bottom left side, the line starts with glass washing. The front side encapsulant sheet is placed on the glass by an operator. Two fully automated stringers deliver the strings on the layup, which is completed manually with the second encapsulant sheet and the backsheet. After lamination and further processing the modules are flashed and unloaded at the top left side into module binning boxes (not visible in this illustration).

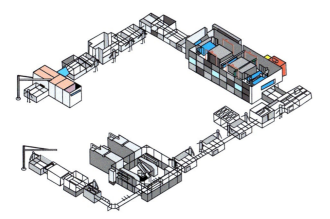

Fig. 3.47: Semiautomated module production line. Reprinted with the permission of SCHMID Group, SCHMID Technology Systems GmbH.

Fully automated cell stringers are operated with a throughput of 2000–4000 cells per hour and line. They remove cell by cell from a stack and use vision systems to check cell integrity, to check the metallization pattern, and to calibrate cell position before placing it precisely on a conveyor belt or in carriers. Anomalous cells are sorted out.

Flux may be introduced by different methods. Some stringers guide the ribbon through a bath with liquid flux (e.g. an isopropyl alcohol solution) and dry it afterwards. The temperature and solid content of this bath have to be controlled continuously. Other stringers spray the flux on the cell's contact pads or on the ribbon. Spraying fluxes the required surfaces selectively, in contrast to dipping. The ribbon side opposite to the cell is not contaminated with residual flux. After evaporation of the solvent, a white powder of flux crystals remains on the ribbon. Some manufacturers provide ribbons that are pre-coated with a flux layer. Cells and ribbons have to be stored carefully to avoid severe oxidation of contact pads and solder surface.

After the ribbon has been cut to the required length which is two times cell length plus cell gap, cells and ribbons are placed in the final position. The ribbons have to link each cell rear side with the front side of the following cell, which enforces a sequential process for common front-back-contact cells. Back-contact cells on the other hand enable a layup with simultaneous placement of two or more cells and a simultaneous interconnection process.

The cell layup then passes the preheating zones of the stringer, where the cells are homogeneously heated. The temperature reaches the flux activation temperature shortly before the actual soldering starts.

The soldering station within the stringer may use very different heating technologies, including contact, infrared radiation, induction, hot air, or laser. Equipment based on these technologies competes in terms of process stability, speed, accuracy, yield, up-time, and ultimately total cost of ownership (TCO). The precise control of the temperature profile from preheating till cooling by pyrometry or other appropriate means is essential for stringer quality assurance. Manufacturers claim to achieve cell breakage rates below 0.2%, depending on throughput, cell thickness, process temperatures, and several other parameters.

Contact soldering requires two contact tools: one for heating and one for holding down the ribbon while the solder is still liquid. The heating tool usually consists of a row of pins that are applied on the top ribbon (Figure 3.48). Soldering process temperature is controlled in the heating pins. The heat is transmitted by contact through the cell to the bottom side ribbon and the solder is molten. The downholder is then applied in between the heating pins to keep the ribbons in place and the heating tool is removed. After solder solidification, the downholder tool is moved back and the transport belt advances the entire string by one cell towards the cooling zone. The heat transfer by contact requires permanently clean contact surfaces of the heating tool.

Infrared soldering uses lamps to heat cell and ribbon by radiation and narrow downholders to minimize shading (Figure 3.49). Pyrometers continuously measure the temperature on the ribbon or on the cell. They deliver the control signal that steers lamp power within each soldering process cycle. In order to receive a meaningful pyrometer signal, infrared reflections within their spectral response range have to be excluded, and the surface emissivity of the measurement spot has to be highly reproducible. While the heat required for soldering is transmitted contactless, the cells are still touched by the downholder.

Challenges for the soldering equipment may arise from discontinuous busbars, where the ribbon has to be soldered to localized contact pads. If the height of the

Fig. 3.48: Soft-Touch soldering tool for 4-busbar-cells with four rows of 11 heating pins in each row; downholder tool is not visible in the heating process step. Reprinted with the permission of Somont.

Fig. 3.49: Cell transport belt for infrared soldering with rows of downholders that apply from both sides to secure the three ribbons in place (left, reprinted with the permission of Teamtechnik) and inductive soldering unit (right, reprinted with the permission of Xcell Automation Inc.) in a cell stringer.

surrounding metallization on the front or rear side exceeds the pad height, soldering is also more difficult. After soldering, the string moves on into a zone for controlled cooling and is finally checked electrically and/or optically by a camera. If accepted, the string is placed face-side-down on the encapsulant sheet that has been prepared on the front cover glass. Figure 3.50 shows a fully automated stringer including the mentioned processing steps.

For string interconnection, the projecting cell ribbon ends are connected by using a dedicated string ribbon with a much larger cross section to minimize resistive losses. The same string ribbons are used to conduct the current to the junction box. After string interconnection, the rear side encapsulant and the backsheet are applied.

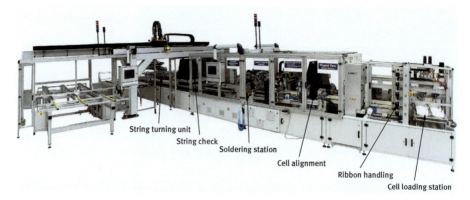

Fig. 3.50: Rapid Two stringer and layup line with process starting at cell loading station (right) and ending with the string layup. Reprinted with the permission of Somont.

The string ribbon ends are fed through openings in these sheets. A final quality check before irrevocable lamination is performed.

The layup then moves on into the laminator, still in face-side-down orientation with the module front glass underneath. Most module production lines use flatbed laminators with a processing chamber divided in two sections by a membrane. The membrane allows independent pressure control in the upper and lower section [72]. In a manufacturing environment, the laminator bottom plate constantly maintains the processing temperature of about 150°C. Many laminators provide an array of pins that avoid immediate contact between the glass pane and the bottom plate at first. Only after the laminator lid has closed and quick evacuation has reduced the chamber pressure to a few millibars, the laminator pins are lowered and the glass is thereby placed on the hot laminator plate. Too quick heating of the glass plate from the lower side causes bow and may displace the cell matrix. The bow originates from the fact that the thermal expansion of the lower glass surface exceeds the thermal expansion of the upper surface. The encapsulant should not melt before complete evacuation, otherwise gas bubbles may remain in the laminate. An additional requirement refers to the dimensional stability of the encapsulant sheet. During the heating phase and before melting, the sheet is supposed to display only limited shrinkage (about 1–2 %) in order to avoid displacement of module components before lamination.

After evacuation and heating, the encapsulant melts. Pressure is then applied through a membrane or a press to distribute the liquid encapsulant around the cell matrix. If EVA is used, thermally activated initiators start the crosslinking reaction to form an elastomeric material. To assure production quality, the laminator bottom plate has to provide a homogeneous temperature distribution over the used area, e.g. in the range of ± 1 °C. Local deviations in temperature impose severe risks on production quality and may cause premature module degradation in the field.

Different technologies are used for heating the base plate. They compete in terms of temperature homogeneity, heating speed, maintenance cost, and other aspects affecting total cost of ownership. Manufacturers offer electric heating, oil heating, hybrid systems combining electric and oil heating, and inductive heating.

Since lamination is a bottleneck in module production, several approaches have been suggested to increase throughput. Many laminators provide enlarged work areas to process a batch of four or more modules side by side, at the cost of an increased equipment footprint within the production line. Some manufacturers provide multi-level laminators as an alternative to the mentioned single-level laminator. With up to 10 levels on top of each other, 20 modules in common sizes can be processed simultaneously. Other manufacturers split the entire lamination process in a prelamination step and a curing step (Fig. 3.51), or provide heating of the laminate from both sides, especially for glass-glass modules.

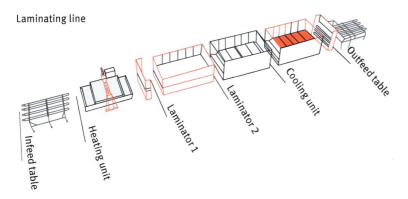

Fig. 3.51: Lamination line. Reprinted with the permission of Meyer Burger.

A cooling press can achieve fast and controlled cooling of the laminate down to room temperature, as compared to passive cooling in ambient air. At the end of the production line, module temperature should be close to 25 °C to allow electrical characterization at standard testing conditions (STC).

After lamination, projecting polymer film needs to be removed from the module edges. In the trimming process, this waste material is removed by cutting. When edges are clean, the frame profiles can be applied through a press. In order to achieve a strong and tight connection, the laminate edge is previously covered by tape on its entire perimeter.

As the last component, the junction box with cables is mounted and connected to the projecting string ribbon ends. The module moves on to the flasher station, where the cables are used for connection. Inline flashers commonly use pulsed xenon lamps as light source and process modules horizontally. If the flasher requires a distance of several meters between light source and receiver plane to achieve sufficient spacial uniformity, the modules are measured face side up, underneath a flasher tower. Some manufacturers provide flasher equipment of less than 1 m in height which achieve grade A spacial uniformity and processes modules inline face side down.

LED flashers offer a compact design, since they require only a small distance between the LED field and the module sample. As light source they use arrays of LEDs with different spectral distributions to synthesise the AM 1.5 spectrum.

Flashers use electronic loads to record a full IV-curve of the module at STC within a few milliseconds flash duration. While electrically connected, an electroluminescence picture of the module is often also taken to check for cell cracks. The measured electric data of the module is printed out on a label in a standardized manner, e.g. according to EN 50380 [73], and attached to the module. Finally, a visual inspection is performed and the modules are sorted into appropriate power classes. Figure 3.52 shows a flashlist for a power bin with 10000 modules where the nameplate power is referenced with 0 %. The module selection is positive, since only modules rated above the nameplate power from 1.2 % to 3 % are included. The fine white horizontal lines stem from rounding of the power value. The decreasing number of modules towards the upper limit power limit at +3 % indicates that this bin is taken from the positive side of the power distribution of the entire production batch.

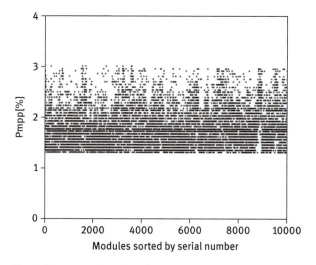

Fig. 3.52: Exemplary flashlist from a module production line including 10000 module power measurements (Fraunhofer ISE).

3.5 Module recycling

Wafer-based PV modules not only contain valuable materials, but in many cases also environmentally hazardous materials. Considering the annual production volume of several tens of GW, which corresponds to a mass of several million tons, module recycling is a must. Modules to be disposed may be rejected from production, they may have been damaged in transit or mounting, or identified as early failures in the field, altogether typically less than 1 % of the total production volume. Aside from these literally virgin modules, deployed modules that approached or exceeded their expected service life of 25–30 years constitute the second important class. These worn-out modules will probably surpass the 1-GW-mark per year in the late 2020s, considering the fact that global installations first climbed above the 1-GW-threshold in the year 2004.

A module recycling process is supposed to separate materials and provide them on a level of purity required for reuse in the same product, or in case of downcycling, for use in less demanding applications.

The amount of the most precious materials contained in PV modules, silicon and silver, has been reduced over time. Average silicon wafer thickness has come down from 400 µm in the 1990s to about 180 µm around the year 2008, where it has remained ever since. Also the polysilicon price has severely decreased from 100–200 US$/kg around the year 2010 to values below 20 US$/kg in the year 2014. Both developments leave the recycling of silicon less attractive than it used to be. Average silver consumption has fallen to about 0.13 g per cell in 2014 [74], from about 0.5 g in the year 2005.

Figure 3.53 shows the predominant materials that constitute a typical PV module. Since a 4 mm front glass thickness is assumed, a total module weight of 23.6 kg

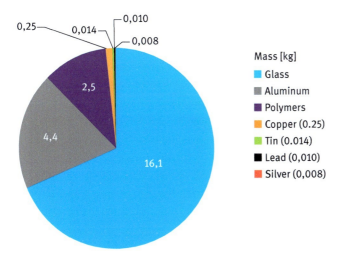

Fig. 3.53: Approximate material quantities included in a typical wafer based module produced in 2014.

is reached. Polymers include encapsulant, backsheet, edge tape, junction box, and cable housing. The copper is contained in the cell ribbons, string ribbons, junction box, and cables. Tin is found in the solder and wire coating. Most of the lead inside the module is contained in the solder, but some lead is also used in the glass additives responsible for firing the metallization paste through the ARC cell coating. Both solder and metallization paste are available in lead-free qualities at a somewhat higher TCO. Aside from lead, fluoropolymers often used as backsheet layers may impose environmental risks. Some solar glass manufacturers use antimony as an additive to improve transmittance.

If the material amounts are weighted with material cost, the potentially most valuable materials pertaining to modules are silicon, aluminum, and silver.

The aluminum frame and the junction box are easily removed. Since the solar cells are very thin (about 180 µm or less), strongly bonded in the laminate and often cracked or broken when modules reach the recycling plant, the wafers cannot be extracted as a whole.

Owing to the difficulties encountered in the separation of the glass-foil laminate, this laminate is shred, which separates the glass from the residual laminate. The broken bits of glass can be introduced into the glass recycling process. The residual laminate containing encapsulant, cells, ribbons, and backsheet is usually burnt in a waste incineration plant, eliminating the organic components. More efficient material separation processes which recover the silver and the silicon directly are under investigation.

In contrast to laminated modules, the glass panes of edge sealed, non-laminated modules (Section 3.2.4) can easily be separated to extract the solar cell strings.

In the European Union, the Directive 2012/19/EU of July 2012 on waste electrical and electronic equipment (WEEE) for the first time explicitly included PV modules. Accordingly national laws had to be effected in the EU countries by February 2014 to ensure that the module producer or another responsible party takes back and recycles modules free of charge. In Germany, the country with the largest installed PV capacity by the end of 2014, a law was prepared in 2015 that requires module manufacturers and distributors to take back and recycle waste modules. The costs are assigned according to the market share of manufacturers and distributors.

4 Basic module characterization

4.1 Light IV measurement

The performance-related electrical properties of PV modules are mainly determined through the indoor measurement of IV-curves using solar simulators. The most important set of conditions is defined as standard test conditions (STC). They require a spacially uniform, orthogonal irradiance of 1000 W/m^2 according to the AM 1.5 spectrum, and a module temperature of 25 °C. The measurement specifications are detailed in different parts of IEC 60904 [7].

The equipment quality in terms of the generated spectrum, uniformity and stability is rated by the letters A/B/C; a triple "A" simulator ("AAA") achieves spectral mismatch below 25 %, spatial nonuniformity below 2 %, and irradiance instability below 2 %, as defined by IEC 60904-9 [75]. High precision laboratory equipment can reduce these uncertainty values by 50 % relative and are commonly labeled "A+A+A+".

The **spatial uniformity** of solar simulators can be assessed by using a specially prepared PV module where every single cell is contacted to the outside. With the help of a multiplexer, the short circuit current of a large number of cells, giving a precise signal for local irradiance, is measured consecutively within milliseconds.

High precision measurements with uncertainties at or below 2 % require spectral correction based on the spectral response of the module under test. This response is separately measured as short circuit current of a single cell or of the entire module using a series of monochromatic filters. It is important to observe the stability of the lamp spectrum in time, since lamp aging usually causes spectral shifts.

At a continuous irradiance of 1000 W/m^2, it is challenging to maintain a constant module temperature. Steady state simulators may take hours to achieve thermal equilibrium. For quick measurements, pulsed sun simulators, so-called flashers, are used (Fig. 4.1). With a flash duration of about 10–25 ms, the module will not heat up appre-

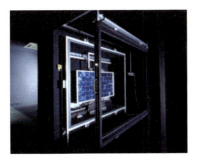 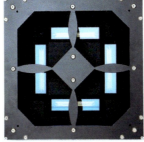

Fig. 4.1: Flasher laboratory at CalLab PV Modules, Fraunhofer ISE, with module mounting cabinet (left) and light source comprising four Xenon lamps (right, reprinted with the permission of Meyer Burger).

ciably. During a period of constant irradiance within the total flash time, electronic loads quickly shift the working point of the module from open voltage to short circuit. In this time, the IV-curve of the module is recorded. To increase measurement precision and to check for capacitance effects, the load curve is passed through in the opposite direction during a second flash. By careful control of the measurement conditions, laboratories can achieve measurement uncertainties for wafer-based modules down to 1.6 % in power. Precisely characterized modules are required as calibration modules for inline flashers in the production lines.

Similarly to the cell case (Section 2.2), the module IV curve leads to a set of particularly important operation points called module IV parameters: the short circuit current I_{SC}, the open circuit voltage V_{OC}, the current, voltage and power at the module's maximum power point (I_{mpp}, V_{mpp}, P_{mpp}) where the IV product reaches its highest value. Fill factor FF and module efficiency are calculated analogously to a single cell (equation (4.1)):

$$FF = \frac{V_{mpp} \cdot I_{mpp}}{V_{OC} \cdot I_{SC}} = \frac{P_{mpp}}{V_{OC} \cdot I_{SC}}, \quad \eta_{mod} = \frac{P_{mpp}}{E_{STC} \cdot A_{mod}}. \tag{4.1}$$

For bifacial modules, separate single-side flasher measurements for the front and the rear side are performed using a black background. This setup rules out current contributions from the opposite side and leads to **front** and **rear single-side efficiency** values η_{front} and η_{rear}. Single measurement characterization of bifacial modules can be obtained either by simultaneous irradiance on both module sides using mirrors or, more suitable for inline control, by adding the rear side irradiance to the front side.

If only STC module IV parameters and temperature coefficients are known, the IV parameters for different irradiance levels or temperatures than STC can be approximated by equation (4.2). This rough approximation is only valid in the vicinity of STC, since it does not include effects related to series and parallel (shunt) resistances:

$$I_{mpp} = I_{mpp,STC} \frac{E}{E_{STC}} \Big(1 + \alpha_{rel} \cdot (T - T_{STC})\Big),$$

$$V_{mpp} = V_{mpp,STC} \frac{\ln(E)}{\ln(E_{STC})} \Big(1 + \beta_{rel} \cdot (T - T_{STC})\Big), \tag{4.2}$$

where

I_{mpp}	current at maximum power point at conditions outside of STC (A),
$I_{mpp,STC}$	current at maximum power point at STC (A),
E	irradiance (W/m²),
E_{STC}	irradiance at STC (1000 W/m²),
T	module temperature [°C],
T_{STC}	module temperature at STC (25 °C),
α_{rel}	module short circuit current relative temperature coefficient [1/°C],
β_{rel}	module open voltage relative temperature coefficient [1/°C].

The entire module can be modelled as a series interconnection of the cell's equivalent circuits (Sections 2.2.1 and 2.2.4). From this module equivalent circuit, a module series resistance R_S can be derived straightforwardly by adding the cell series resistances and introducing additional series resistances that originate from cell interconnection and cables. R_S does not correspond to localized ohmic components in the module, but rather captures the effect of many distributed parasitic resistances at a defined operating point of the module. The series resistance of a module thus comprises ohmic effects from the sheet resistance of the cell emitter, from the cell metallization, the cell and string interconnectors, and the module connection cables, which show a positive temperature coefficient, the bulk resistances of the wafers with their negative temperature coefficient, and the cell junction resistances. Quality issues in module production, breakage, and corrosion may increase the series resistance, which in turn reduces module fill factor and power. In consequence, the series resistance determination provides a diagnostic tool for quality assurance and degradation assessment.

Different procedures for determining module series resistance from measured light IV curves have been proposed [76]. A simple estimation for R_S can be derived from the slope of the IV curve at the open circuit voltage (equation (4.3), [77]). Best approximations are achieved for IV curves recorded at high irradiance levels. This simple estimation is useful for assessing changes in R_S rather than its absolute value:

$$R_S = -\frac{dV}{dI}\bigg|_{I=0}. \tag{4.3}$$

The standard in IEC 60891 [78] describes a procedure to transform an IV curve (I_1, V_1) measured at a defined irradiance E_1 and temperature T_1 to a second IV curve (I_2, V_2) that corresponds to a second set of parameters E_2 and T_2. In equations (4.4) and (4.5), the relative temperature coefficients of current and voltage are denoted with α_{rel} and β_{rel}:

$$I_2 = I_1 + I_{SC,1} \cdot \left[\left(\frac{E_2}{E_1} - 1\right) + \alpha_{rel} \cdot (T_2 - T_1)\right], \tag{4.4}$$

$$V_2 = V_1 - R_S \cdot (I_2 - I_1) - \kappa \cdot I_2 \cdot (T_2 - T_1) + V_{OC,1} \cdot \beta_{rel} \cdot (T_2 - T_1). \tag{4.5}$$

The procedure suggests to measure IV curves of a module at constant temperature and three or more different irradiance levels. R_S is then fitted to a value that reaches best agreement when all measured IV curves are transformed to the IV curve recorded at maximum irradiance. Prior to this fitting procedure, the curve correction factor κ has to be determined from a set of IV curves recorded at the same irradiance and different temperatures [78]. This is only necessary if the IV curve transformation includes a change in temperature.

The parallel resistance (shunt resistance) of a cell can be approximated by the negative slope of the voltage with respect to the current at $I = I_{SC}$ (equation (4.6), Section 2.2.3). If this slope is read from a module IV curve and divided by the number of cells in a module, an approximate mean parallel cell resistance is gained:

$$R_P = -\frac{dV}{dI}\bigg|_{V=0}. \tag{4.6}$$

The parallel resistance of single cells within a module can be measured destructively by directly contacting the cell through openings in the backsheet. Nondestructive methods have also been suggested, which rely on measuring light IV curves of the module while shading single cells [79].

Before characterization, PV modules have to be preconditioned to reach a stable state. The preconditioning assures measurement reproducibility and is usually also meant to reach a relevant condition for outdoor operation. It is well known that thin film devices may exhibit reversible or irreversible changes in their properties, depending on their operation or even storage history. In wafer-based silicon PV modules, most notably light-induced degradation may lead to a onetime power drop. Figure 4.2 shows a constant light simulator and the LID behavior of a module. Only after about six hours, which corresponds here to about $22\,\text{MJ/m}^2$ ($6\,\text{kWh/m}^2$) irradiation, a stable state is reached.

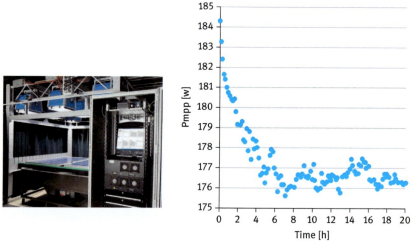

Fig. 4.2: Steady-state irradiance simulator (left), exemplary modules degradation over time of a poly-Si module exposed to a constant irradiance of 1000 W/m² (Fraunhofer ISE).

4.2 Energy rating

In outdoor operation, modules encounter various conditions with respect to temperature, irradiance intensity, irradiance distribution, and spectrum. For a precise yield prediction, additional flasher measurements are performed under different conditions than STC, especially at elevated temperatures and low irradiance. Figure 4.3 shows a set of measurements according to the energy rating standard IEC 61853 [80]. While all modules exhibit a negative temperature coefficient in power, some modules may show an efficiency increase around 800 W/m², before the positive irradiance coefficient in voltage begins to dominate at lower irradiance levels and reduces efficiency.

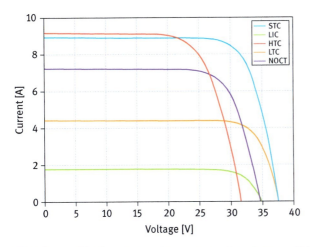

	G	T	Isc	Voc	Impp	Umpp	Pmpp	η	
	[W/m²]	[°C]	[A]	[V]	[A]	[V]	[W]	[%]	condition
STC	1000	25	8.94	37.6	8.39	30.4	255.1	15.3	Standard test
LIC	200	25	1.77	34.9	1.66	29.4	48.8	14.6	Low irradiance
HTC	1000	75	9.12	31.6	8.35	24.1	201.2	12.1	High temperature
LTC	500	15	4.24	37.6	4.18	31.6	132.0	15.8	Low temperature
NOCT	800	47	7.24	34.6	6.72	27.7	186.2	14.0	Nominal operating cell temperature

Fig. 4.3: IV-curves and module parameters determined in an energy rating measurement (Fraunhofer ISE).

For bifacial modules, additional measurements with (simultaneous) bifacial irradiance are performed to approach field conditions. The front and rear single side efficiency values only give an approximation for bifacial operation efficiency, due to nonlinear loss and gain mechanisms related for example to series resistance and V_{OC}. Bifacial operation may be simulated by providing a diffusely reflecting background. A more precise simulation is obtained by arranging a pair of mirrors to simultaneously irradiate both module sides from one light source or by using two synchronized flasher.

4.3 Dark IV measurement

Dark current-voltage measurements on photovoltaic strings and modules provide valuable information for production quality assurance and for degradation monitoring [81]. While dark IV curve measurements give no information on the light-induced short circuit current of a module, several parameters of the 2-diode-model can be observed.

When dark IV is measured, a current driven by an external source flows through the solar cells in forward bias (right side of Figure 4.4). The light and the dark IV curves are very similar, except for the shift originating from the light induced current (Figure 2.3) and for a lower series resistance effective in dark IV measurement. This difference originates from the different current paths. If the solar cell is operated under light as a generator (left side of Figure 4.4), most of the charge carriers need to travel additional lateral paths to the front electrode. If the cell is operated as a forward biased diode (right side of Figure 4.4), carriers travel directly in between the electrodes where they encounter less series resistance.

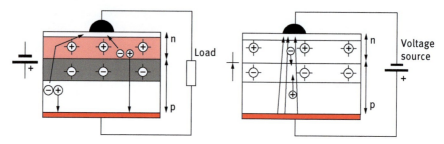

Fig. 4.4: Illuminated solar cell in operation (left) and solar cell under dark condition operated as a diode in forward bias (right).

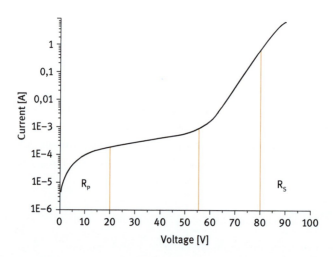

Fig. 4.5: Sketch of a module dark IV-curve indicating sensitivity regions for different parameters of the 2-diode solar cell model (Fraunhofer ISE).

On the light IV curve, I_{SC}, V_{OC} and the maximum power point (V_{mp}, I_{mp}) are readily identified. If the dark IV curve is shifted to this I_{SC} along the current axis, a hypothetical open circuit voltage and maximum power point can be assigned to the dark IV curve.

A semilog plot is more usefull for analyzing dark IV characteristics of PV modules. Figure 4.5 shows where different model parameters of the 2-diode-model (Section 2.2.4) influence the IV curve segments. Reduced parallel resistance R_P due to shunting will increase the current in the left segment at low voltages. The series resistance determines the curve behavior at large currents, and increasing R_S reduces the current on the right side of the plot.

4.4 Electroluminescence imaging

Solar cells are designed to generate free charge carriers by using the energy of absorbed photons. This process can be reversed to a certain extent. If an external voltage is applied to a solar cell, charge carriers will recombine. A small fraction of these recombination processes are radiative, leading to an infrared emission that peaks around 1150 nm [82]. The generated local electroluminescence (EL) radiant exitance $M_e(x)$ depends on the local voltage across the cell, the temperature voltage and a calibration factor according to equation (4.7) [83]:

$$M_e(x) = C(x) \cdot \exp\left(\frac{V(x)}{V_T}\right) \quad \text{if} \quad V(x) \gg V_T, \tag{4.7}$$

where
$M_e(x)$ local electroluminescence radiant exitance [W/m^2],
$V(x)$ local voltage [V],
V_T temperature voltage [V],
$C(x)$ calibration factor.

The local voltage for a cell spot x can be estimated by correcting the entire cell voltage for voltage drops which originate from the series resistance in the particular current path to the spot x. This resistance is determined by the path length along the metallization and the emitter sheet. Defects in solar cells related to the crystal structure (grain boundaries, lattice dislocations), to precipitated impurities as well as cell cracks locally reduce radiative recombination and appear dark.

For EL imaging, cameras based on silicon CCD arrays or InGaAs photodiode arrays are available. CCD cameras provide high resolution, but their signal-to-noise ratio suffers from the fact that the spectral response of the silicon CCD array overlaps only sparsely with the EL emission spectrum. InGaAs photodiode arrays show a better spectral response, but provide less resolution.

EL images of a PV module can reveal comprehensive information on the mechanical integrity and partly on the electric functionality of the solar cells. The fault pattern helps to identify the origin of the fault, which may be attributed to silicon crystallization, cell manufacturing, or module manufacturing, as far as production is concerned.

Totally dark cells in between active cells in a module are shunted. Completely dark regions within the active area of a solar cell indicate complete disconnection, which is usually caused by interruption of the front- or rear-side metallization. In field or laboratory aged modules, chemically driven degradation (corrosion) may also severely increase the contact resistance between the finger and the wafer or reduce finger cross sections, which also leads to dark regions. Corrosion tends to result in a gradual transition from dark to lighter areas, whereas breakage leads to abrupt changes in the EL signal.

Completely dark regions caused by cracks occur more often between an outer busbar and the cell edge, since these regions are only linked to one busbar. Regions between busbars are contacted to busbars from two sides. Due to this redundancy, they only appear completely dark when both connections are interrupted. Some cracks

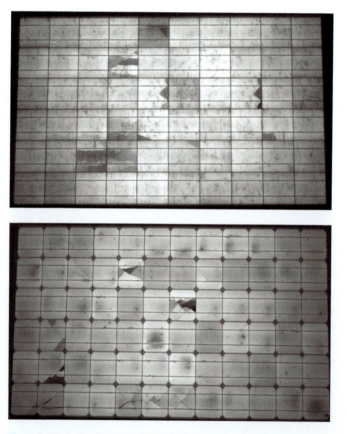

Fig. 4.6: Electroluminescence images of a 60-cell poly-Si (top) and a 72-cell mono-Si module with several faults recognizable as dark lines and dark regions within cells (Fraunhofer ISE).

disconnect the metallization only partially, allowing a reduced current flow. EL signal intensity decreases along the fingers from the busbar to the point at maximum distance from the busbar or from both adjacent busbars. If a crack interrupts the finger, the distance to this point may increase, leading to a distinct, gradually increasing darkening.

The top image in Figure 4.6 shows EL images of poly-Si module. In these cells, lattice dislocations and grain boundaries make it more difficult to detect cracks, especially when they only partially disconnect metallization. In mono-Si solar cells, cracks tend to follow crystal axes, which often leads to diagonal paths and cross-like crack shapes (bottom image in Figure 4.6). A monocrystalline wafer without lattice dislocations and grain boundaries facilitates the detection of cracks.

5 Module power and efficiency

5.1 IV parameters and electric model

A PV module can be regarded as a mainly two-stage transformation of the used cells. First, cells are interconnected in series to form a cell string. Then this string is encapsulated. The electric properties of a cell string can be characterized by an IV curve similar to the IV curve of single solar cells. A similar set of IV parameters can be derived from the string's IV curve: I_{SC}, V_{OC}, I_{mpp}, V_{mpp}, P_{mpp}, fill factor FF and efficiency η.

As for the electric model, the cell string can be regarded as a series interconnection of the 1- or 2-diode cell models, with some additional series resistance originating from ohmic losses due to current collection and transportation in between cells. These losses can be taken into account by correcting the series resistance R_S of each cell in the cell model, leading to somewhat different I_{mpp}, V_{mpp}, P_{mpp}, fill factor FF and efficiency η.

In consequence, we obtain the string IV curve by a superposition of the individual corrected cell IV-curves: at every current value, the string voltage corresponds to the sum of the corrected cell voltages. Figure 5.1 shows the principle in case of identical cells. The IV curve of one cell displays a reverse breakdown voltage of −14 V and a short circuit current of 9.5 A at 1000 W/m² irradiance.

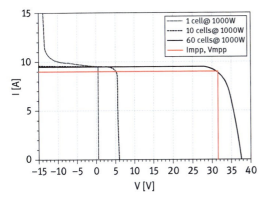

Fig. 5.1: Calculated IV curves for one cell, 10 cells and 60 cells in series connection at 1000 W/m² irradiance; maximum power point values of current (I_{mpp}) and voltage (V_{mpp}) are indicated.

The string power at each current value is the sum of the corrected power values over all cells at this current. The current and voltage mpp values marked red in Figure 5.1 result in a string mpp power ($P_{mpp} = I_{mpp} * V_{mpp}$) of 282 W.

If all cells have identical properties, the string mpp current will be identical with the corrected single cell mpp current; the same applies to the string's short circuit current and its fill factor. The efficiency referring to the total active cell area is also conserved in this case, provided that stringing does not change the irradiance on the active area, e.g. by additional shading caused by interconnectros.

If the string is encapsulated, effective irradiance is changed for all cells in a similar manner due to various optical effects. Since the light induced current is affected, encapsulation mainly changes I_{SC}, I_{mpp} and P_{mpp}.

In practice, a set of solar cells that are connected to a string will display some small variation in I_{mpp}. As a result, string I_{mpp} will slightly differ from individual cell I_{mpp}, leading to an operation of these cells slightly outside of their maximum power point. String mpp power will be somewhat lower than the sum of all cell mpp power (Section 5.6).

5.2 Partial shading and hot-spots

What happens if inhomogeneous irradiance or deviating cell performance leads to severe variations in cell short circuit current? Let us assume that one cell only receives a mean irradiance of 200 W/m² due to partial shading of the string, while all other cells receive 1000 W/m². As shown in Figure 5.2, the weak cell only contributes to string voltage in the low current range. After string current exceeds the short circuit current of the weak cell, this cell is operated in reverse bias, progressively reducing the string's total voltage. Only after exceeding the breakdown voltage of the shaded cell, the string current would recover. Yet, this is not desirable, since the shaded cell would heat up severely.

The mpp power of the configuration in Figure 5.2 only amounts to 154 W. Although the shading affects only 80 % of one cell out of 60 cells, which amounts to 1.3 % irra-

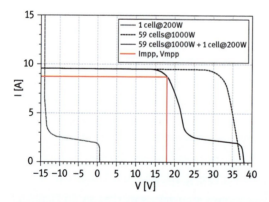

Fig. 5.2: Calculated IV curves for one cell at 200 W/m² irradiance, 59 cells in series connection at 1000 W/m² irradiance, and for all 60 cells in series connection; maximum power point values of current (I_{mpp}) and voltage (V_{mpp}) are indicated.

diance reduction for the entire string, the string power is reduced by more than 45 % compared to the fully illuminated case.

Aside from this overproportional power loss, heat generation in the partially shaded solar cell is a severe issue. The missing power, in this case 128 W, is dissipated in the partially shaded cell. This dissipation may occur in spacial inhomogeneous patterns, leading to extreme concentrations of current and heat generation densities in some spots. These hot-spots may destroy polymeric module materials (encapsulant, backsheet) and the solar cell itself, and may set materials on fire.

Due to these power losses and irreversible damage risks, solar cells require protection which limits their maximum reverse bias voltage to values below their specified breakdown voltage. One option would be to equip every single cell with a protective diode. This diode has to be connected reversely to the solar cell diode such that it becomes conductive whenever the reverse bias voltage on the solar cell exceeds the threshold voltage of the protective diode. This measure limits the string's maximal power loss and prevents hot-spots on solar cell, given that the protective diode can safely dissipate enough power.

In practice, cell integrated protective diodes have proven too expensive for most applications. For common 60 cell modules therefore a compromise is chosen by dividing the entire string into typically three substrings. Each substring is bypassed through a diode which becomes conductive in case of negative voltage bias on this substring (Figure 3.42).

If the diode becomes conductive, the top substring does not deliver any more power to the external load, leading to a power loss of the entire module in the range of 1/3. Yet, this loss is lower than in case of Figure 5.2.

Figure 5.3 displays the IV curves for this partial shading condition with the diode. As soon as the short circuit current of the weak substring is exceeded, the bypass diode

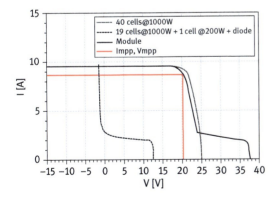

Fig. 5.3: Calculated IV curves for two fully operating substrings with 20 cells each, for a 20 cell substring with one partially shaded cell and a bypass diode, and for the entire series connection of the three substrings; maximum power point values of current (I_{mpp}) and voltage (V_{mpp}) are indicated.

takes over. The resulting module mpp power reaches 176 W, which corresponds to 62 % of the fully illuminated case.

As a second consequence of the substring bypass diode, only the voltage of the 19 cells in full operation is applied as a reverse bias voltage to the partially shaded cell. As long as this voltage in the range of −10 V is well below the breakdown voltage of each single cell, no critical hot-spot will occur. For safe operation, the breakdown voltage specification of the cells needs to be higher than the maximum reverse bias voltage originating from the residual substring. Cells with a lower breakdown voltage thus require smaller substrings and more bypass diodes.

5.3 Power and efficiency model

Module efficiency is mainly governed by solar cell efficiency, but approximately 10–15 % relative are controlled by the way solar cells are interconnected and encapsulated. There are three types of effects that influence the efficiency change from cell to module: geometrical effects, optical effects, and electrical effects.

The nominal power of a module is determined to a large extent by the power of its active parts, the cells. We therefore define as starting point P_0, the sum of the total nominal (STC) power of all n cells that compose the module (equation (5.1)). The nominal power of each cell at its individual mpp, $P_{cell,j}$ is determined in a flasher measurement in ambient air, prior to encapsulation:

$$P_0 = \sum_{j=1}^{n} P_{cell,j}. \tag{5.1}$$

When these cells are integrated into a module with front cover, encapsulant, backsheet, and interconnectors, the effective irradiance on the active cell area changes. These changes may have a variety of origins: new or changed optical interfaces, absorption from cover materials, shading by interconnectors, multiple reflections, and scattering. As a result, the short circuit current of the cells will change, together with the entire IV-curve.

If the irradiance dependency of the nominal cell power is linear in the vicinity of the STC point, which is normally the case and can be checked by flasher measurements at different irradiance levels close to 1000 W/m^2, cell mpp power will change similarly to the short circuit current. It follows that optical changes will proportionally affect power as long as they apply to all cells in a similar manner.

Aside from these optical effects due to packaging, electrical effects arise from serial interconnection of the cells. When cells are connected in a series, they always operate at a common current. This current is usually set by an mpp tracking device at a point where the cell string, the entire module, or a string of several modules delivers maximum power. If the mpp current of single cells within the module varies, not all cells can operate at their individual mpp. As a consequence, string mpp power will be less than the sum of individual mpp power of all cells. This effect can be captured in a mismatch loss factor that depends on the variance of the mpp cell currents and therefore can be controlled by cell sorting.

A second electric effect arises from series resistance losses due to current transport inbetween cells, and usually also due to current collection and transport to the cell edge. The ohmic resistance of the interconnection circuit will cause voltage drops and power loss proportional to the second power of the current.

From these considerations, nominal module power P_{mod} (always at STC in the mpp) can be expressed as the cell power sum P_0, corrected by several **change factors** f_i originating from the optical and electrical effects mentioned [84, 85]:

$$P_{mod} = P_0 \cdot \prod_{i=1}^{m} f_i. \tag{5.2}$$

Change may mean loss or gain; losses lead to change factors smaller than 1. Electrical change factors are always smaller than unity, since the cell flasher measurement provides optimal conditions for eliminating series resistance losses. Optical change factors may affect module power in both directions. If only j change factors are considered instead of all m factors, the module power at this intermediate step P_j is expressed as

$$P_j = P_0 \cdot \prod_{i=1}^{j} f_i. \tag{5.3}$$

If the effects are considered to be independent, it is possible to attribute a module power change to a certain effect, e.g. absorption loss in the encapsulant or resistive loss across the cell gap:

$$\Delta P_j = P_j - P_{j-1} = P_0 \cdot \prod_{i=1}^{j-1} f_i \cdot (f_j - 1). \tag{5.4}$$

According to this definition, losses with their corresponding change factors smaller than 1 result in negative values for the corresponding ΔP_j. Many effects are interdependent. If for example cell spacing in a module with white backsheet is increased, a higher module current and a higher series resistance due to increased ribbon length will follow. Hence, the individual power loss terms need to be interpreted with care.

With the above definitions, the final module power may also be expressed as a sum of power changes that apply to the initial cell power (equation (5.5)):

$$P_{mod} = P_0 + \Delta P_1 + \Delta P_2 + \ldots + \Delta P_m. \tag{5.5}$$

Nominal efficiency of a PV module is defined as the ratio between the delivered electrical power at the maximum power point (P_{mpp}) and the irradiance on the full module area (A_{mod}) at standard test conditions (STC) with $E_{STC} = 1000\,W/m^2$ (equation (5.6)):

$$\eta_{mod} = \frac{P_{mod}}{E_{STC} \cdot A_{mod}}. \tag{5.6}$$

The full module area is always larger than the total cell area. The solar cells require a minimum distance from the module edge for electrical insulation and safety requirements. Between the cells, gaps are required to avoid a short circuit and to allow for ribbon transition from front to back. This totally or partially inactive area generates efficiency losses, because it increases the nominal area A_{mod}.

Since an inactive module area contributes little or nothing to module efficiency, and this area is disproportionally large for small module samples, the scientific community sometimes uses alternative definitions to characterize laboratory module samples or prototypes. Very common is "aperture area efficiency", which considers in A_{mod} only the area covered by solar cells, sometimes expanded by 1–2 mm of inactive area around each cell, if optical gains are expected from there.

The module efficiency η_{mod} can also be expressed as a sum based on the initial module efficiency η_0, corrected by efficiency changes η_j that may be attributed to different physical effects along the way from cell to module (equation (5.7)):

$$\eta_{mod} = \frac{P_0 + \Delta P_1 + \Delta P_2 + \ldots + \Delta P_m}{E_{STC} \cdot A_{mod}} = \frac{P_0}{E_{STC} \cdot A_{mod}} + \sum_{j=1}^{m} \frac{\Delta P_j}{E_{STC} \cdot A_{mod}} = \eta_0 + \sum_{j=1}^{m} \eta_j. \tag{5.7}$$

Equation (5.7) defines an initial module efficiency η_0 and efficiency changes η_j. The latter corresponds to the mentioned power changes and change factors from optical and electrical effects. The difference between η_0 and the mean cell efficiency η_{cell} originates from the efficiency loss due to the area increment from total cell area to module area, i.e. the addition of inactive area. This loss $\Delta\eta$ is defined in equation (5.8):

$$\eta_0 = \eta_{cell} + \Delta\eta. \tag{5.8}$$

The module area A_{mod} can be divided into the sum of the cell areas, $n * A_{cell}$, the total area in between cells, A_{gap}, and the total border area of the module, A_{border} (equation (5.9)):

$$A_{mod} = nA_{cell} + A_{gap} + A_{border}. \tag{5.9}$$

In case the solar cells are shingled (Section 3.1.4), A_{gap} should be regarded as a negative value corresponding to the entire cell area covered by cell overlap.

With the distinction in module area, efficiency losses can be attributed to the cell gaps and the module border (equation (5.10)):

$$\Delta\eta = \eta_0 - \eta_{cell} = \frac{P_0}{E_{STC} \cdot A_{mod}} - \eta_{cell} = \eta_{cell}\left(\frac{nA_{cell}}{A_{mod}} - 1\right) = -\eta_{cell}\left(\frac{A_{border}}{A_{mod}} + \frac{A_{gap}}{A_{mod}}\right). \tag{5.10}$$

Similarly to the power change factors, efficiency change factors can be defined (equation (5.11)). But unlike the power change factors, the efficiency change factors are zero if no loss or gain occurs:

$$f_{border} = -\frac{A_{border}}{A_{mod}}; \quad f_{gap} = -\frac{A_{gap}}{A_{mod}}. \quad (5.11)$$

The geometrical efficiency loss due to inactive area can be expressed as

$$\Delta\eta = \eta_{cell}f_{border} + \eta_{cell}f_{gap}. \quad (5.12)$$

The module efficiency can be expressed based on mean cell efficiency with distinct geometrical, optical and electrical effects as

$$\eta_{mod} = \eta_{cell} + \eta_{cell}f_{border} + \eta_{cell}f_{gap} + \sum_{j=1}^{m}\eta_j; \quad \eta_j = \frac{\Delta P_j}{E_{STC} \cdot A_{mod}}. \quad (5.13)$$

The inactive areas also affect the terms η_j in equation (5.13), but since $\Delta P_j \ll P_0$, these effects are of the second order and will be neglected in the discussions to follow. Table 5.1 gives an overview on the change factors f_j which will be discussed in the following sections.

Tab. 5.1: Change factors (loss or gain) for module efficiency and power with respect to the cells.

Change factor	Type	Main responsible effects
f_{border}	geometrical	inactive border area
f_{gap}		inactive cell gaps or overlap for shingled cells
f_1	optical	air/glass interface reflection
f_2		glass bulk absorption
f_3		glass/encapsulant interface reflection
f_4		encapsulant bulk absorption
f_5		active area interface reflection
f_6		active area multiple reflection (involving the cover material/air-interface)
f_7		cell finger multiple reflection
f_8		cell interconnector shading and multiple reflection (involving the cover material /air-interface)
f_9		cell spacing multiple reflection (involving the cover material /air-interface)
f_{10}	electrical	cell missmatch
f_{11}		cell stringing series resistance
f_{12}		series resistance of string interconnector
f_{13}		series resistance of module cables and plugs

5.4 Geometrical effects

The distance between live parts (including the entire cell matrix) and the module metal frame is required to be 16 mm according to IEC 61730-1 [86] for system voltages up to 1000 V. The frame itself may add 2 mm on each side, giving a total minimum border with of 18 mm. Each string interconnector with a typical width of 5 mm and a spacing of 2 mm may require 7 mm in total if it is not placed behind the cell matrix. Common junction boxes require additional space, e.g. 14 mm. In this example, there are two borders with 18 mm, one border with 25 mm, and one border with 39 mm that make up the inactive border area. If 156 mm cells with 2 mm gaps are used, the module length would result to 1642 mm and module width to 982 mm. In consequence, inactive area would amount to 9.4 % of the total module area, with $f_{border} = -7.4\%$ and $f_{gap} = -2\%$. The border fraction can be reduced by increasing the number of cells per module, e.g. going from 60 to 72 cells.

Mono-Si solar cells are often produced in a pseudo-square format instead of full square. Pseudo-square format helps to reduce material waste when cylindrical mono-Si ingots are cut to square wafers, at the expense of cell packing density in the module. Figure 5.4 shows common situations with mono-Si wafers of an 8 inch and 6 inch format. In the aforementioned example with 60 cells in the 156-mm-format (6 inch), f_{gap} losses increase from −2 % to −3.2 % due to the missing corners. A white backsheet or rear side encapsulant helps to recover part of the cell gap efficiency losses (Section 5.5.7).

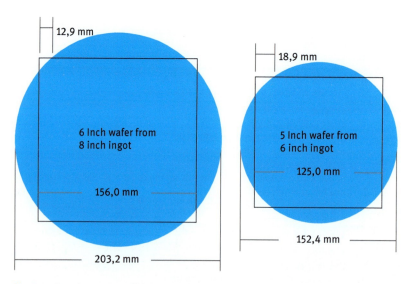

Fig. 5.4: Pseude-square cell formats originating from mono-Si ingots.

If a central junction box is not necessary because the bypass diodes are integrated into the module and module cables are attached to opposite module edges (Figure 3.38), inactive module border area can be reduced by 0.5 % absolute or more.

5.5 Optical effects

Module power is affected by the spectral properties of the front side layers and their interaction with the cell surface, the irradiance spectrum and the internal spectral response of the solar cell. For a first assessment of these layers, effective values need to be extracted from the spectral values over the decisive wavelength range. Weighting is performed with the solar spectrum, because transmittance values at wavelengths with high solar spectral density are more important than those where solar irradiance is low. This weighting leads to the solar transmittance and is relevant for solar thermal collectors where the conversion efficiency is largely independent of the wavelength over a broad spectral range. In contrast, solar cells display a quantum efficiency that declines quickly around their bandgap at 1100 nm. Towards the UV range (Figure 2.12), the conversion efficiency declines, since only a decreasing fraction of the photon energy can be used. The blue light response depends on cell technology, which means that different cells may require different encapsulation materials for optimum performance.

Before encapsulation, the short circuit current of a solar cell, $I_{SC,ref}$, can be expressed by using the spectral response (Section 2.4) according to equation (2.33). If a semitransparent layer (interface or sheet) is placed in front of the solar cell and we assume that this layer only acts as a homogeneous filter with no multiple reflections or any other interactions with the cell, the layer will reduce the short circuit current to $I_{SC,layer}$. Its effective transmittance, T_{eff}, can be expresses according to equation (5.14):

$$T_{eff} = \frac{I_{SC,layer}}{I_{SC,ref}} = \frac{\int_{300\,nm}^{1200\,nm} T(\lambda) \cdot SR(\lambda) \cdot E_\lambda d\lambda}{\int_{300\,nm}^{1200\,nm} SR(\lambda) \cdot E_\lambda d\lambda}. \qquad (5.14)$$

According to its definition, the external spectral response SR includes reflectance losses at the interface air/ARC/wafer. This interface is modified by encapsulation, usually reducing losses at the new interface encapsulant/ARC/wafer. Simulation shows that for an ideal random pyramid texture with SiN_x ARC coating, the reflectance for frontside air of about 2 % is reduced to about 1 % for frontside encapsulant. For a more precise assessment of the effective transmittance of a layer, the SR used in equation (5.14) can be therefore corrected to reproduce the encapsulated situation. The correction may be derived from ray tracing simulations and will increase the SR particularly in the UV and IR range.

In the encapsulated module, different interfaces and materials are placed in front of the cell: an air/glass interface possibly equipped with an ARC, the glass bulk, the glass/encapsulant interface and the encapsulant bulk. While T_{eff} can be calculated for each individual interface and bulk material from the stack, the product of these effective transmittances is not physically meaningful. It will only give an approximate value for the effective transmittance of the entire stack, since each layer usually changes the irradiance spectrum that reaches the subsequent layers. The precise value of T_{eff} for a stack of materials has to be calculated by inserting the spectral stack transmittance in equation (5.14). The fact that layers change the spectrum received by subsequent layers also has to be considered when T_{eff} is calculated for the second, third, or fourth layer by using the actually transmitted spectral irradiance.

5.5.1 Air/glass and glass/encapsulant interface reflection (f_1, f_3)

Within the relevant part of the solar spectrum, soda-lime glass shows normal dispersion, meaning that the refractive index decreases with wavelength, and an effective refractive index of 1.52. Figure 5.5 displays spectral refractive indices for a choice of transparent materials.

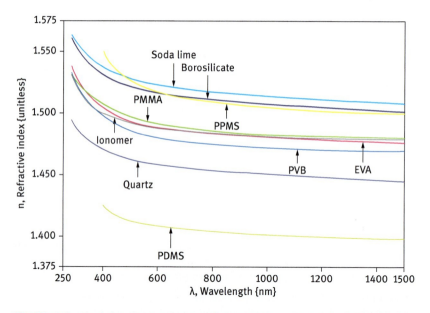

Fig. 5.5: Refractive index of various cover and encapsulation materials for PV modules. Image from [87].

The angle-dependent reflectance for a flat, uncoated air/glass interface is given by the Fresnel equations. For normal incidence, these equations lead to a simple relationship for the reflectance R and the transmittance T of the interface, depending on the two refractive indeces involved, and independent of the polarization (equation (5.15)):

$$R = \left(\frac{n_2 - n_1}{n_2 + n_1}\right)^2 \quad T = 1 - R. \tag{5.15}$$

When the dispersion n(λ) is known, the effective transmittance T_{eff} of the air/glass interface can be calculated using equation (5.14). When the dispersion is not known, interface transmittance T and reflectance R can be calculated from spectral measurements of a glass slab transmittance and reflectance. In such a transmittance measurement, light needs to pass the air/glass interface, the bulk material, and the glass/air interface.

Figure 5.6 shows transmittance results based on measurements on glass samples. The effective transmittances for the uncoated glass interface results to 95.8 % and for the AR-coated glass interface to 99.0 %. These effective transmittance values correspond to the change factor f_1.

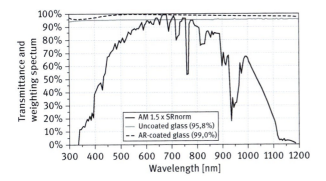

Fig. 5.6: Transmittance of the air/glass interface for uncoated and AR-coated solar glass derived from measurements and weighting spectrum; the figures in parenthesis indicate effective transmittance values.

The transmittance of this AR coating drops in the blue range, coupled with an increased reflectance, which leads to a blueish appearance of the coated glass.

Equation (5.15) does not apply for AR coated glass, where R needs to be determined by coherent reflectance simulation, if the ARC parameters are known, or by reflectance measurements. Neither does equation (5.15) apply for heavily structured glass surfaces which lead to multiple reflections (AR structures, Section 3.2.1). In these cases, ray-tracing simulations are required to assess transmittance.

The small refractive index difference between glass (approx. n = 1.52) and common encapsulants (n = 1.45...1.5) leads to a negligible interface reflectance; for silicon-based materials with their refractive index in the range of 1.4...1.42, the glass/silicon interface reflectance reaches 0.2 %, which is still low compared to other loss effects. In consequence, the change factor f_3 can usually be set to 1. In special cases where no encapsulant is used in between glass cover and cell matrix, a gas will be the next optical medium behind the glass, with a refractive index of 1. The reflectance of such an interface depends on the AR-treatment of the rear surface of the front glass.

5.5.2 Glass and encapsulant bulk absorption (f_2, f_4)

Impurities in the glass (Section 3.2.1), the encapsulant, and its additives absorb a small part of the incoming irradiance [88]. When monochromatic beam radiation passes through a nonscattering, homogeneous medium, it is attenuated depending on the path length and on the absorption coefficient of the medium (equation (5.16)). The bulk transmittance T_{bulk}, defined as the ratio of the irradiance after a path length d, E(d), with respect to the initial irradiance E(0) is given by

$$T_{bulk} = \frac{E(d)}{E(0)} = \exp(-\alpha \cdot d), \qquad (5.16)$$

where

T_{bulk}	bulk transmittance,
E	irradiance [W/m^2],
α	absorption coefficient [1/m],
d	path length [m].

According to the definition, bulk transmittance does not comprise any interface effects. The absorption coefficient is a material property and varies with the wavelength. Figure 5.7 shows the bulk transmittance spectra of a float glass in common architectural (green) grade compared to a solar grade glass.

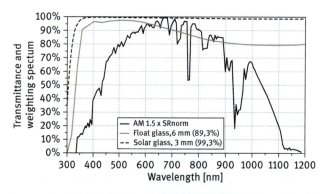

Fig. 5.7: Bulk transmittance of different glass qualities and weighting spectrum; the figures in parenthesis indicate effective transmittance values.

The effective bulk transmittance $T_{bulk,eff}$ of the cover glass corresponds to the change factor f_2. It is calculated in the same manner as the effective transmittance T_{eff} in equation (5.14) by weighting with the AM 1.5 spectrum and the cell spectral response. The data displayed in Figure 5.7 leads to an effective bulk transmittance of 89.3 % for the green glass and 99.3 % for the solar glass.

Figure 5.8 shows bulk transmittance values for different encapsulants. The effective bulk transmittance $T_{bulk,eff}$ of the encapsulant corresponds to the change factor f_4. The silicon material displays a very high UV transmittance, it seems to contain no UV-blockers. The ionomer by contrast absorbs a large part of the UV radiation. The EVA and the polyolefin behave very similar in the UV range, yet the polyolefin shows a slightly higher absorptance towards short wavelengths.

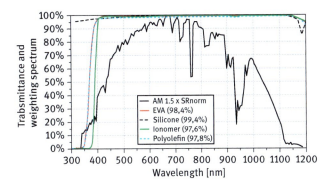

Fig. 5.8: Bulk transmittance of different encapsulants and weighting spectrum, with the effective bulk transmittance values inside the parentheses.

When bulk transmittance effects are analyzed by means of short circuit measurements in module flashers, the UV range of the lamp spectrum deserves special attention. Only those lamps can be used which reproduce the AM 1.5 short wave length irradiance.

5.5.3 Active area interface reflection (f_5)

Changes in the active area interface reflectance originate from the substitution of air as the ambient medium by an encapsulant, changing the reflectance of the material stack medium/AR coating/silicon. Ideally, the refractive index of a single layer AR coating is chosen such that $n_{encap}/n_{AR} = n_{AR}/n_{Si}$, inside the relevant wavelength range (Figure 2.14).

For $n_{encap} = 1.48$ and $n_{Si} = 3.9$, this would require $n_{AR} = 2.4$. If such a cell is characterized in ambient air before encapsulation, the interface reflectances on both sides of the AR coating are no longer balanced, and the AR coating would not work properly. In this case, after lamination there would be a current gain of the cell due to a reduction of the active area interface reflectance. In practice there are limitations to use a single-layer AR coating based on the prevalent SiN_x with such a high refractive index. Instead, these coatings typically display refractive indeces of about 2.1, which is in between the optimal values for an air and an encapsulant environment.

Figure 5.9 shows the spectral coherent reflectance calculations for a SiNx AR coating of 75 nm thickness on a flat silicon substrate for ambient air and ambient EVA. The calculations have been performed with the OPAL online tool from pvlighthouse.com.au and consider coherent wave interaction and absoptance in the ARC.

The effective reflectance values obtained in this case are quite similar, despite the curve shape difference, which leads to a change factor f_5 close to unity.

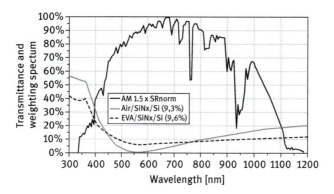

Fig. 5.9: Spectral reflectance of different material stacks on a flat silicon wafer.

If the silicon surface is textured, double or even triple reflections occur within the surface. As a result, the effective reflectance drops significantly (Section 2.10).

If the AR coating is deposited on a plane wafer, f_5 can be deducted by comparing reflectance measurements before and after encapsulation at relevant angles. Measurements after encapsulation have to be corrected for parasitic effects like the reflectance of the encapsulant/air interface. Relevant angles are defined by the texture surface profile.

If the AR coating properties (refractive index, thickness) and the texture surface profile (especially the predominant local incidence angles) are known, f_5 can also be calculated using a coherent reflectance model.

5.5.4 Active area multiple reflection (f_6)

Solar cells are structured in order to reduce reflection losses at the front surface (Section 2.10). Cell encapsulation also changes the interaction of the texture with light by introducing an additional reflection.

For normally incident light, rays penetrate the cover and the encapsulant and impinge on the cell texture at the same angle than before encapsulation. Rays reflected from the textured cell surface will usually impinge on the glass/air interface at oblique incidence. Depending on this angle, the glass/air interface will partially or totally reflect the light back to the solar cell.

Figure 5.10 shows a schematic cross section of an ideal texture on a mono-Si solar cell. The pyramid facets are inclined at 54.7° with respect to the cell plane, a fixed value due to crystal orientation. The incident ray (0) is reflected twice, and the secondary reflection (2) is lost in the initial state (left side). After encapsulation, a third reflection (3) redirects light partially to the cell. For refractive index values above 1.6, the third reflection would be a total internal reflection and most of the light reflected by the cell would be redirected to it. Yet, for the common encapsulant EVA with a refractive index of 1.48, the reflectance for ray (2) is only about 15 %. Reflection losses from normally incident light can thus be partially recovered due to the interaction of cell texture and glass/air interface, especially for encapsulants with higher refractive index.

Real textured surfaces, especially on poly-Si cells, generate reflections at different angles. If the angular distribution of the reflected light is measured on the active cell area, the effect of encapsulation and the corresponding change factor f_6 can be derived. If the cell surface profile can be modelled, a ray tracing simulation can also help to quantify this recovery.

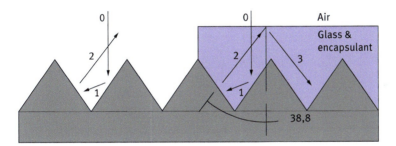

Fig. 5.10: Ray paths on a textured mono-Si solar cell for normally incident light before (left) and after (right) encapsulation.

For obliquely incident rays, light refraction at the air/glass interface reduces the effective incidence angle at the cell surface. A ray incident at 60° with respect to the glass will be refracted to impinge at about 35° into the cell plane. This incidence angle reduction will usually increase the efficiency of the cell's AR treatment. Associated gains become relevant in non-STC conditions (Section 6.4), but not for the determination of nominal module power at normal incidence.

5.5.5 Finger multiple reflection (f_7)

Metallization fingers cover 2–4 % of the active cell area. Before encapsulation, the shading effect from these fingers corresponds to their geometrical width. Light incident on their bright silver surface is mostly lost, since it is diffusely reflected into the front side hemisphere, or, to a small extent, absorbed. After encapsulation, backscattered light is partially reflected at the glass/air interface, depending on the incidence angle. Reflected light will usually reach the active cell area and contribute to current generation. The additional reflection effectively reduces the optical shading loss. We therefore introduce an effective finger width $w_{F,eff}$ that corresponds to the width of a hypothetical black finger which causes the same shading loss than the real finger in the encapsulated state. The corresponding change factor f_7 is given in equation (5.17). Before encapsulation, $w_{F,eff} = w_F$, leading to $f_7 = 1$.

$$f_7 = 1 + \frac{A_F}{A_{cell} - A_F} \cdot \left(1 - \frac{w_{F,eff}}{w_F}\right) \tag{5.17}$$

where
- A_F cell area covered by fingers [cm²],
- A_{cell} total cell area [cm²],
- $w_{F,eff}$ effective finger width [µm],
- w_F geometrical finger width [µm].

In the encapsulated state, the light backscatterd from the fingers travels inside a medium with refractive index of about 1.5 and encounters the interface to air with a refractive index of 1. According to Fresnel's equations, the reflectance of this interface increases from about 4 % at normal incidence to 100 % for angles larger than the total internal reflection angle (α_{TIR}) given in equation (5.18). For EVA with $n_{enc} = 1.48$, α_{TIR} amounts to 42.5°:

$$\sin \alpha_{TIR} = \frac{1}{n_{enc}}. \tag{5.18}$$

If Lambertian scattering is assumed, which leads to isotropic radiance L_e at all observation angles, the portion of the reflected radiant flux that encounters TIR is given by equation (5.19) [89]:

$$\frac{\Phi_{e,TIR}}{\Phi_e} = \frac{\int_{\alpha_{TIR}}^{90°} L_e \cdot \cos(\vartheta) \cdot \sin(\vartheta)\, d\vartheta}{\int_0^{90°} L_e \cdot \cos(\vartheta) \cdot \sin(\vartheta)\, d\vartheta} = [\cos(\alpha_{TIR})]^2 = 1 - \frac{1}{n_{enc}^2}, \tag{5.19}$$

where
- $\Phi_{e,TIR}$ reflected radiant flux at angles exceeding ϑ_{TIR} [W],
- Φ_e total reflected radiant flux [W],
- α_{TIR} TIR angle [°],
- L_e reflected radiance [W/m²/sr],
- n_{enc} refractive index of encapsulant.

Rays that are diffusely reflected at angles smaller than α_{TIR} are only recovered to a small extent. If they are neglected, we obtain a simple relationship between the relative effective width, finger reflectance and refractive index (equation (5.23)). For black fingers ($R_F = 0$) or ambient air ($n_{enc} = 1$), the effective width is identical with the geometrical width:

$$\frac{w_{F,eff}}{w_F} = 1 - \left(1 - \frac{1}{n_{enc}^2}\right) \cdot R_F. \tag{5.20}$$

The diffuse reflectance of a silver finger surface, R_F, usually amounts to about 90 %. For the common encapsulant EVA with $n_{enc} = 1.48$, this results in an effective finger width in the encapsulated state which amounts to about 50 % of the geometrical width. In this case, f_7 would amount to 1.01…1.02 for finger coverage ratios of 2…4 %.

According to equation (5.20), an encapsulant with a lower refractive index, for instance a silicone with $n_{encaps} = 1.4$, will result in a reduced TIR range when compared to an EVA with $n_{encaps} = 1.48$ [89]. Such an index reduction results in a 10 % relative reduction of the achievable gain from scattering, not only from cell fingers, but also from diffusely reflecting ribbons (Section 3.1.1) or cell spacing (Section 5.5.7). By contrast, an encapsulant with higher refractive index further increases current gains from backscattered irradiance. The current gains from multiple reflections only depend on the refractive index of the encapsulant that contacts the scattering surface; the index of the front glass or other cover materials, even an AR coating on the front glass does not affect these gains.

5.5.6 Cell interconnector shading and multiple reflection (f_8)

When common ribbons are applied to the active side of a solar cell without covering the active cell area, cell current will not change due to stringing. Yet, in order to save silver, the metallization area is often smaller than the geometrical width of the interconnector. The busbars are often tapered towards the cell edge, since solder joints will not be positioned too close to the edges. In other cases, continuous busbars are replaced by discontinuous contact pads (Section 2.9) or totally omitted. When these reduced or shaped busbars are covered by ribbons, formerly active areas that contributed to cell current will be shaded. This results in current losses after string formation, but still before encapsulation, in an air environment. The change factor $f_{8,air}$ can easily be determined by comparing inactive area before and after interconnection (equation (5.21)). Losses may also occur when the ribbon is not placed precisely on top of the busbar (missalignement losses):

$$f_{8,air} = \frac{L_{cell} - n_{ribbon} \cdot w_{ribbon}}{L_{cell} - n_{ribbon} \cdot w_{ribbon}(1 - f_{act})} \tag{5.21}$$

where

n_{ribbon} number of ribbons on active cell side,
w_{ribbon} ribbon width (diameter) [mm],
L_{cell} cell edge length perpendicular to ribbon direction [mm],
f_{act} fraction of the projected ribbon area that covers active cell area.

If the cell has no busbars or contact pads at all and the ribbon contacts the fingers directly, f_{act} would be very close to unity. In the other extreme, if the cell has continuous busbars that are precisely covered by the ribbons, f_{act} amounts to zero. In case that the solar cells are shingled (Section 3.1.4), n_{ribbon} equals one and w_{ribbon} is equal to the overlap width.

Interconnectors can also contribute positively to cell current by means of single reflections. If interconnectors provide a nonrectangular cross section, they can reflect part of the light that impinges within their geometrical width towards active cell area. This is the case for round wires (Figure 5.11A), or it would occur with trapezoidal or triangular wire cross sections. The current gains from these interconnectors would raise the change factor f_8, possibly to values above unity, depending on the specular surface reflectance. The effective width of these interconnectors is smaller than their geometrical width, in analogy to the effective and geometrical finger width in the previous chapter.

The situation for a round wire is shown in Figure 5.11. Rays incident orthogonally in section A are redirected from the wire to the active cell area, independently of the encapsulation.

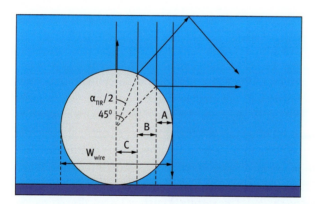

Fig. 5.11: Schematic cross section of an encapsulated round wire interconnector with specularly reflecting coating resting on a solar cell; light reflection of different sections (A,B,C) is displayed with limiting ray paths.

Due to these reflections, the effective width of the wire is smaller than the geometrical width w_{wire}. The change factor f_8 for a number of wires per cell, n_{wire}, and a cell edge length L_{cell} can be estimated according to equation (5.22) in air (before encapsulation). Only for a hypothetical black wire does the surface reflectance R_{wire} become 0 and the effective width of the wire equals its geometrical width. Equation (5.22) describes the situation after string formation and before encapsulation:

$$f_{8,air} = \frac{L_{cell} - n_{wire} \cdot W_{wire,eff}}{L_{cell} - n_{wire} \cdot W_{wire}(1 - f_{act})}; \quad W_{wire,eff} = W_{wire}\left[1 - R_{wire}(1 - \sin(45°))\right] \quad (5.22)$$

where

n_{wire} number of wires on active cell side,
W_{wire} wire width (diameter) [mm],
L_{cell} cell edge length [mm],
f_{act} fraction of the projected wire area that covers active cell area,
R_{wire} effective reflectance of wire surface (considering the AM 1.5 spectrum and the cell's spectral response).

In the encapsulated state, additional current gains may occur from interconnectors due to secondary reflections at the glass/air interface, especially for rays that encounter TIR. The entire change resulting from shading and multiple reflections is captured by the change factor f_8. Only for common rectangular ribbons with flat, specularly reflecting surfaces does the situation not change after encapsulation, and the final change factor f_8 equals $f_{8,air}$.

In the case of round wires, additional gains from secondary reflections further reduce the effective width of the wire after encapsulation. The silver cell metallization and bright interconnection wires or ribbons reflect back a substantial portion of the incident light. In Figure 5.11, the rays incident in section B are redirected from the wire to the glass/air interface and totally reflected to the active cell area. Only rays reflected from section C mostly leave the module. In this encapsulated state, the change factor can be estimated according to equation (5.23):

$$f_8 = \frac{L_{cell} - n_{wire} \cdot W_{wire,eff}}{L_{cell} - n_{wire} \cdot W_{wire}(1 - f_{act})}; \quad W_{wire,eff} = W_{wire}\left[1 - R_{wire}\left(1 - \sin\left(\frac{\alpha_{TIR}}{2}\right)\right)\right] \quad (5.23)$$

The saw-tooth profiled ribbon with specular nonsymmetric reflection in Figure 3.6 is supposed to redirect all normally incident rays to the active cell area by total internal reflection at the glass/air interface. In practice, nonideal profiles that scatter some light outside of the TIR angle range and coating reflectance values below 100 % still lead to losses. Equation (5.24) provides estimations for the change factor f_8. With silver-coated, saw-tooth profiled ribbons, the effective ribbon width can be reduced to about one third of the geometrical ribbon width in the encapsulated state:

$$f_8 = \frac{L_{cell} - n_{ribbon} \cdot W_{ribbon,eff}}{L_{cell} - n_{ribbon} \cdot W_{ribbon}(1 - f_{act})}; \quad W_{ribbon,eff} = W_{ribbon}\left[1 - R_{ribbon,TIR}\right] \quad (5.24)$$

where

$R_{ribbon,TIR}$ effective reflectance of wire surface into the TIR angle range (considering the AM 1.5 spectrum and the cell's spectral response).

The diffusely reflecting ribbon in Figure 3.6 distributes rays over the entire hemisphere. Those rays that hit the glass/air interface at angles larger than α_{TIR} will be totally reflected towards the cell area. The corresponding effect can be estimated in a similar way than in case of light scattering fingers (Section 5.5.5) and leads to equation (5.25). The scattering coating only becomes effective in the encapsulated state. For a hypothetical black ribbon (ribbon reflectance $R_{ribbon} = 0$) or an air environment with a refractive index n_{enc} of 1, only shading losses remain. In a realistic situation, a white, scattering ribbon coating showing 85–90 % diffuse reflectance reduces effective ribbon width by about 50 %:

$$w_{ribbon,eff} = w_{ribbon}\left[1 - R_{ribbon}\left(1 - \frac{1}{n_{enc}^2}\right)\right] \tag{5.25}$$

5.5.7 Cell-spacing multiple reflection (f_9)

Most PV modules are assembled with a white backsheet or a white rear-side encapsulant sheet. Light incident in the cell gaps, in the missing corners of pseudo-square cells, and next to the frame is diffusely reflected (backscattered), depending on the material reflectance. Backscattered rays that hit the glass/air interface at an angle larger than the critical TIR angle are redirected back into the module. Under the simplifying assumptions of equation (5.19), about half of the backscattered radiation is captured by TIR, depending on the encapsulant's index of refraction. If it reaches the active cell area, it will increase the short circuit current. Ray tracing simulation is required to calculate the optical gains resulting from this backscattering.

The first parameter that controls these current gains directly is the effective reflectance of the rear side material. A black backsheet does not contribute at all, at least in the visible range of the spectrum. High backsheet reflectance values in the range of 90 % maximize the current gain.

The second parameter is the cell spacing. Larger cell gaps mean large backscattering areas. If the backscattered light is totally internally reflected and reaches the active cell area, a current gain will result. The missing corners of the pseudo-square cells also act as backscattering areas. If cell distance is getting very large, however, an increasing portion of the backscattered light will miss the active cell area and become backscattered once again.

The third parameter is the ratio between cell perimeter and cell area. Five-inch cell formats or divided cells in particular will profit more from cell spacing multiple reflections than common six-inch cells.

The fourth parameter is the distance between the backscattering layer (usually the backsheet) and the cell front side. A white rear-side encapsulant layer is more effective than a transparent rear-side encapsulant combined with a white backsheet. In the latter case, some backscattered rays will hit the rear side of the solar cells or cell edges. These rays are lost as long as the module is built up with monofacial solar cells. In contrast, a white rear-side encapsulant located close to the cell front surface plane will direct a vast portion of the backscattered rays towards the glass/air interface without the aforementioned losses.

Figure 5.12 shows the measured current gains on one-cell-samples where the cell border variation has been implemented by increasing aperture sizes. At 0 mm, the aperture has the same size as the solar cell. The current gain saturates with increasing border width, since backscattered light will increasingly impinge on the backsheet a second time. In this measurement setup, the current gain may be slightly higher than in a multicell module with corresponding cell spacing. This is due to contributions from the shaded part of the back scattering material.

Cell-spacing multiple reflection is useful for increasing module power, but the partial losses in the inactive area reduces module efficiency, which is referenced to the entire area. Larger cell spacing or the use of divided cells will thus increase all area-related module costs like material cost for glass, encapsulant, backsheet on the module side, and system cost associated for example to mounting and land use.

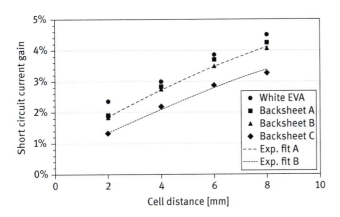

Fig. 5.12: Measured current gains for different materials and cell border widths; the dashed lines are exponentially fitted curves. Data from [84].

Rays that are backscattered at angles smaller than the critical TIR angle are mostly transmitted and leave the module. To avoid these losses, saw-tooth profiled, specularly reflecting films with high reflectance, similar to Figure 3.6, have been applied to the cell spacing areas. Figure 3.32 shows a photograph of such a module sample. For an ideal structure and reflectance, all normally incident rays are thereby reflected towards the glass/air interface at angles above the TIR angle. From this interface, the rays are totally reflected towards the cell active area. In reality, there will occur some losses due to rounded edges of the profile and to a reflectance smaller than 100 %. For oblique incidence, part of the reflected rays miss the total reflectance condition and escape to the ambient.

The change factor f_9 associated with cell spacing multiple reflection corresponds to the current gain displayed in Figure 5.12, increased by 1. For modules build-ups without encapsulation, where there is a gas layer in between the front cover and the solar cell, no backscattered light will experience TIR. In this case, the reflectance of the back cover has hardly any influence on the cell current and the change factor can be assumed to be 1.

5.6 Electrical effects

5.6.1 Cell mismatch (f_{10})

After production, solar cells are sorted into bins according to their maximum power, which corresponds to an efficiency **binning**. For the assembly of one module, only cells from one efficiency bin are normally used. In the module, these cells are operating in a series circuit at a common current $I_{mpp,mod}$. The change factor f_{10} due to mismatch losses (equation (5.26)) can be expressed as the sum of the power of all cells operated at the common module mpp current, $I_{mpp,mod}$, divided by the sum of the power of all cells at their individual mpp current, $I_{mpp,i}$. By this definition, f_{10} is always smaller or equal to one:

$$f_{10} = \frac{\sum_{i=1}^{n} P_{cell,i}(I_{mpp,mod})}{\sum_{i=1}^{n} P_{cell,i}(I_{mpp,i})}. \tag{5.26}$$

If the IV curves of all cells are measured previous to module assembly, and the module I_{mpp} is measured afterwards, both numerator and denominator of f_{10} can be calculated. The mismatch loss in a module depends on the distribution width of the I_{mpp} values in the cell batch and on the sensitivity, more precisely on the slope of the cell power as a function of the current at the mpp point. A narrow distribution and a flat power maximum around I_{mpp} will lead to low mismatch losses. The diode model used in Section 2.2 behaves quite stable around the maximum power point: a 1 % change in current around I_{mpp} only leads to power losses in the range of 0.1 %.

The I_{mpp} distribution of cells within a power bin may be approximated by a normal distribution or a section within a normal distribution. It is convenient to use the standard deviation to describe the spread of the values. Statistical investigations [90] have shown that the cells that entered module production in the investigated case show a standard deviation in their I_{mpp} in the range of 0.008 within one power bin. The resulting mismatch effect on module power is below the detection limit. On the other hand, for module prototypes beyond series production where the cells may show a significant spread, it is important to consider mismatch loss.

5.6.2 Series resistance losses (f_{11}, f_{12}, f_{13})

The reference measurement for the assessment of series resistance losses from cell to module is again the cell characterization performed in a flasher. A conventional screen printed solar cell is placed on a metal chuck and thereby electrically contacted on its full back surface. On the front side, rows of current and voltage probes are pressed to the silver busbars, in places designated for the application of interconnection ribbons.

This measurement leads to almost negligible current transport in the designed string direction and generates hardly any associated series resistance losses.

In the connected state of the solar cell, the interconnectors in combination with the cell metallization have to collect the current from the cell area and conduct it to the edge where it can be passed on to the next cell. On the other side of the cell, the current is received at the opposite edge and needs to be distributed over the cell area. Figure 3.4 shows this schematic current flow for the front side metallization.

Allthough the total current is similar, the different current paths give rise to additional series resistance losses.

On the front side, the cell fingers run perpendicular to the string current direction, and hence they cannot contribute to this current. The interconnector ribbon or wire is conducting the cell current from the associated fingers which linearly increases from the start to the maximum value reached at the cell edge. For this reason, the cell length only contributes by a factor of 1/3 to the stringing related power loss on the cell front side. Equation (5.27) gives this power loss for the case of a linear, uniform current increase along the interconnector with constant cross section A_{IC}. This condition applies for a continuous busbar and joint, e.g. a solder line along the entire cell:

$$P_{str,front} = n_{BB} \cdot \frac{\rho_{IC}}{A_{IC}} \left(\frac{L_{cell}}{3} + w_{gap} \right) \cdot \left(\frac{I_{mpp}}{n_{BB}} \right)^2, \tag{5.27}$$

with

$P_{str,front}$	stringing related power loss on cell front side [W],
n_{BB}	number of busbars per cell, equal to number on interconnectors per cell side,
ρ_{IC}	effective electrical resistivity of interconnector material [Ωm],
A_{IC}	cross-section of one interconnector [m²],
L_{cell}	cell edge length in string direction [m],
w_{gap}	cell gap width [m],
I_{mpp}	cell mpp current [A].

The interconnector solder coating contributes to current transport along the interconnector and should therefore be considered. If the interconnectors are not soldered, but glued, e.g. with conductive adhesives, it is necessary to also consider resistive losses associated with the contact and the adhesive material.

If the cell provides continuous busbars as contact pads, they will slightly contribute to lateral conductivity in string direction and thus somewhat reduce the stringing related series resistance losses (typically 10 % or less). This contribution can easily be assessed by measuring their specific resistance (per length unit) and consider them in parallel circuit with the interconnectors.

Increasing the cell gaps (w_{gap}), which is sometimes used as a means to increase the transparency of glass-glass modules in BIPV applications, has detrimental effects on the power for conventional designs.

Since the cell current affects the losses at the power of two, measures that reduce current and increase voltage in a string will effectively reduce the losses. One way to achieve this are divided cells (Section 2.9). Cells can be cut in two or more sections along the string direction, leading to 50 % or more savings in series resistance losses. Divided cells in full-size modules require different interconnection schemes than common square cells due to the reverse bias and hot-spot risk. If a common module is built from three cell strings of 20 cells each, protected by three diodes, a module with half cells would require six cell strings of 20 cells each, protected by six diodes, or a scheme including parallel cell interconnection.

If the number of busbars and interconnectors is increased without changing the total metal cross section of the interconnectors, no benefit will be obtained for the stringing-related resistance losses (equation (5.27)). On the other hand, the losses in the solar cell metallization will decrease, since the maximum current accumulating in the fingers is reduced (equation (2.35)). This concept is important for the multiwire stringing approaches.

On the rear side of common screen printed solar cells, the full area sintered aluminum and Al-Si eutectic layers supports the lateral current flow in string direction substantially. The stringing related power loss on the cell rear side will be overestimated by equation (5.27). A realistic assessment of the effective loss in this 2-dimensional setting can be obtained by the numerically finite element method (FEM) similar to the right side of Figure 3.9. For a sheet resistance of 10 mΩ, (Section 2.9.1) the resistance of the rear-side metallization over the full cell width of 156 mm is comparable to the resistance of a single copper ribbon with a cross section of 1.3 mm × 0.18 mm.

If the entire cell matrix in a module is considered, additional series resistance losses arise at the beginning or end of a string, where the high current needs to reach the string interconnectors. The string interconnectors themselves and the module cables (4 mm^2) are usually chosen with a sufficient cross section to avoid noteworthy losses.

The loss factor f_{12} can be expressed as

$$f_{11} = 1 - \frac{P_{str,front} + P_{str,back}}{P_{mpp}}. \qquad (5.28)$$

f_{11} may account for 3–4 % losses, depending on cell current and total cross section. For elevated operating temperatures, ohmic losses rise with the temperature coefficient of the resistivity, and in the case of copper, silver, and aluminum this amounts to about 0.4 %/°C.

While total interconnector width and thereby cross section is usually limited by shading drawbacks, the cross section alone is also limited by the risk of excessive thermomechanical stress on the single joint.

The remaining factors f_{12} and f_{13} are easily derived by considering the ohmic behavior of the ribbon and wire. Module cables and plugs usually cause small series resistance losses at (or even below) 0.3 %, which would correspond to a change factor value of $f_{13} = 0.997$.

5.7 Comprehensive model

The change factors discussed so far can be displayed within a waterfall diagram showing the stepwise efficiency losses and gains from cell to module (CTM). For an exemplary calculation, the cell and module data in Table 5.2 is assumed.

Tab. 5.2: CTM parameters.

Cell		
edge length	156	mm
P_{mpp}	4.7	W
I_{mpp}	8.66	A
efficiency	19.3	%
busbars	3	
Module design		
cell distance	2	mm
border width (left, right, top, bottom)	39, 25, 18, 18	mm
module length (in string direction)	1642.0	mm
module width	982.0	mm
module area	1.612	m²
module inactive area (cell gaps)	0.033	m²
module inactive area (borders)	0.120	m²
Module materials		
ribbon width	1.4	mm
ribbon height	180	μm
effective interface reflectance of AR coated glass	1	%
encapsulant, effective bulk absorptance	2.1	%
backsheet, effective reflectance	79	%

The inactive module area is responsible for a severe efficiency loss of 1.82 % absolute (Figure 5.13). These losses do not affect module power, which is increased by the cell spacing multiple reflection due to the white backsheet. This gain partly compensates the efficiency loss due to the inactive gap area.

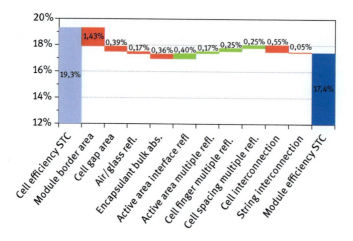

Fig. 5.13: Exemplary calculated waterfall chart for cell-to-module efficiency changes.

The low efficiency loss at the air/glass interface of 0.17 % absolute is only obtained by using an effective AR coating on the glass surface.

The bulk absorptance loss in the encapsulant of 0.36 % absolute is caused by UV blocking additives. Intrinsically more stable materials would allow for a reduction of these additives. Gains are achieved due to the better index matching in the encapsulated state and to several effects relying on total internal reflectance of back scattered light at the glass/air interface. Electrical losses of 0.55 % absolute arise from the cross-section ribbons, whereas the string interconnectors hardly contribute any losses.

The final module has a nominal efficiency of 17.4 %, which indicates a CTM efficiency loss of 1.9 % absolute. The CTM power loss, which does not consider the first two effects of geometrical nature, only amounts to 0.5 % relative in this example. In general, the CTM power loss for poly-Si cells may even turn into a power gain, since these cells display particularly high active area interface reflection gains.

The cost for solar cells and modules are mainly related to their nominal power (W_p), but there also is a distinct bonus for efficiency. More efficient cells help to reduce all area-related and unit costs related to the number of cells to be processed in module production. If total module cost consists of 60 % cell cost and 40 % module-related cost, and the cells are substituted by an improved generation showing 1 % relative higher efficiency and price, module efficiency and power increase by approx. 1 % relative, but module cost would only increase by 0.6 % relative. The same consideration applies to the next step, the PV power plant. If the plant cost comprises 60 % module cost, and the modules are replaced by an improved generation showing 1 % higher efficiency at 0.6 % higher price, and we assume for the sake of simplicity that the same inverters can be used, and the plant power would increase by 1 %, but its cost would only increase by 0.36 %. In this example, a 1 % relative efficiency increase on cell level could justify a cell price increase of 2.8 %.

Similar considerations can be applied on efficiency improvements on the module level. In the above example, the module-related cost amounts to 24 % of the PV power plant cost. If an efficiency improvement of 1 % can be obtained by measures on the module level, this corresponds to a 4 % increase in value added on the module assembly level.

6 Module performance

Up to this point, we focused on nominal module power and efficiency as determined in a flasher measurement under standard test conditions (STC). In **outdoor operation**, PV modules interact with the environment and experience a variety of operating conditions. Along with these varying conditions, the actual efficiency of the module changes, meaning the ratio of electric power output to the irradiance in the module plane. The changes observed with respect to the nominal efficiency are expressed in the so-called **performance ratio** (PR_{mod}) of the module. A PR_{mod} of 100 % would imply that the module permanently operates at nominal efficiency. PR_{mod} can be assessed for any point in time, but usually it is calculated for a certain time period as the ratio of the total generated electric energy to the irradiation in the module plane, referenced to the module STC efficiency (equation (6.1)):

$$PR_{mod} = \frac{E_{DC}}{H_{POA} A_{mod} \eta_{STC}}, \qquad (6.1)$$

with

PR_{mod} (annual) module performance ratio,
E_{DC} (annual) sum of delivered electric energy, assuming mpp tracking operation [MJ],
H_{POA} (annual) irradiation into plane of the module (array) [MJ/m²],
A_{mod} total module area [m²],
η_{STC} STC module efficiency.

It is important to note that PR_{mod} depends not only on module properties, but also on meteorological and mounting conditions, namely orientation, tilt angle, and ventilation. **Stationary mounted** modules receive maximum irradiance if they are oriented towards the equator and tilted at an angle inbetween the geographical latitude and the solar zenith angle in the sunniest season of the year (the module tilt angle references the horizontal, while the solar zenith angle references the vertical). This optimal orientation will also maximize PR_{mod}. **Tracker-mounted** modules follow the apparent path of the sun in the sky. The one- or two-axis tracking systems orientate the module towards the sun, trying to maintain low incidence angles for direct irradiance. PR_{mod} is different for tracker-mounted modules, since they operate at different module temperature levels and irradiance incidence angles.

PR_{mod} can be measured in the field or calculated for a given weather data set, defined mounting conditions, and a comprehensive module property data set. The module data set is acquired following the energy rating standard IEC 61853 [80] as mentioned in Section 4.2.

Based on the data sets, models are required to calculate the POA irradiance, the module temperature, the effective angle of incidence, and for bifacial modules also the rear-side irradiance. The results can be used to rate the module performance with respect to different climates in its initial state. For life time performance prediction, additional assumptions on the annual degradation rates are required, which in turn may again depend on the climate. Usually module manufacturers guarantee limits for long-term power degradation.

The actual AC performance ratio achieved by an entire PV power plant, PR_{plant} (equation (6.2)), includes not only effects from non-STC module operation, but also losses due to (partial) shading, soiling, degradation, electrical losses in the string cables, and due to module mismatch, losses in inverters and transformer, also component or system break-downs and related down time that occurred in the period under consideration:

$$PR_{plant} = \frac{E_{AC}}{H_{POA} A_{mod} \eta_{STC}}, \qquad (6.2)$$

with

PR_{plant}	(annual) plant performance ratio,
E_{AC}	(annual) sum of delivered electric energy, assuming mpp tracking operation [MJ],
H_{POA}	(annual) irradiation into plane of the module (array) [MJ/m^2],
A_{mod}	total module area [m^2],
η_{STC}	STC module efficiency.

External **shading** may originate from obstacles like mountains, buildings, or trees, which reduce the true horizon to the visible horizon. Internal (row) shading is introduced due to space limitations which restrict row spacing. Systems with trackers can avoid row shading if they are able to switch from the regular tracking into the **backtracking** mode where modules do not face the sun any more.

Soiling losses depend on local aerosol concentration and properties, wind incidence, atmospheric humidity, salinity and dew incidence, rain quantity and frequency, module inclination and front cover surface, and module resilience against partial shading. In moderate climates like in central Europe, given module tilt angle larger than about 10 °C and in absence of specific local soiling risks like close foliage trees or construction sites, frequent rain usually reduces soiling losses to a few percent. In contrast, regular module cleaning may be required in arid regions to avoid heavy double-digit losses.

Inverters provide effective efficiencies of about 95–99 %, with large (central) inverters working somewhat more efficient than string or module inverters.

6.1 Irradiation models

The term irradiance refers to power received per unit area and is measured in W/m^2. When irradiance is integrated over a specific time period which may be annual, monthly, daily, or hourly, we obtain the respective **irradiation** on the receiver surface with the unit MJ/m^2. The traditional unit of 1 kWh corresponds to 3.6 MJ. The orientation of the receiver area may be **horizontal, normal** with respect to solar coordinates, or user defined (**plane of the array**, POA).

The **global horizontal irradiance** $E_{glob,hor}$ (or GHI) incident on a horizontal receiver surface can be divided into a **beam** (or "**direct normal**") component and a **diffuse** component. The beam irradiance E_{beam} is usually measured within a view angle of 5.7° width, which is centered at the sun's position and includes circumsolar irradiance. The **diffuse sky irradiance** E_{diff} has been scattered in the atmosphere outside of the beam view angle towards the receiver surface. Tilted receivers also receive **ground-reflected irradiance** E_{ground}.

For the precise mean **annual yield** simulation of a PV power plant in a specific location, long-term local meteorological data covering at least 10 years with a resolution of one hour or better is required. This data must provide mean global horizontal and global diffuse irradiation as well as temperature and wind speed for each time interval. The data can be generated for any location in the world from satellite data calibrated by ground measurements. Data quality depends on the location proximity to ground measurement stations and on local climatic factors.

For yield estimations and performance ratio studies that are less specific in terms of precise location and absolute yield, but intended for the performance comparison of different module types, more averaged and compact data formats are available. A commonly used format is a representative, yet artificially compiled annual data set, a so called **typical meteorological year** (TMY). It provides hourly resolved data for a region and reproduces mean values of the relevant meteorological parameters.

Modelling of **solar coordinates** delivers the angular information for the transformation of horizontal or beam irradiance into POA irradiance. The direction of beam irradiance at a specific location, day, and time is expressed in a spherical coordinate system according to equation (6.3):

$$\sin(\alpha_S) = \cos(\phi)\cos(\delta)\cos(\omega) + \sin(\phi)\sin(\delta)$$

$$\gamma_S = \text{sign}(\omega) \cdot \left| \arccos\left(\frac{\sin(\alpha_S)\sin(\phi) - \sin(\delta)}{\cos(\alpha_S)\cos(\phi)} \right) \right| , \quad (6.3)$$

where

- α_S solar altitude angle (0° ≤ α_S ≤ 90°, 90° is the zenith),
- γ_S solar azimuth angle between south and the beam projection (−180° ≤ γ_S ≤ 180°, west is positive),
- ϕ geographical latitude of the location (−90° ≤ ϕ ≤ 90°, north is positive),
- ω hour angle between the local meridian and the solar noon meridian (−180° ≤ ω ≤ 180°; afternoon, Latin "post meridiem", p.m. is positive),
- δ declination angle (−23.45 ≤ δ ≤ 23.45, north is positive).

The **hour angle** ω captures the angular displacement of the sun east or west of the local meridian due to the rotation of the earth on its axis at 15° per hour. It can be related to local time by considering the local meridian and the equation of time [91]. For performance evaluation, this relationship is less important, and the hour angle is only expressed through **local solar time** (LST, equation (6.4)):

$$\omega = (\text{LST} - 12\text{h}) \cdot 15°/\text{h}. \tag{6.4}$$

The **declination angle** δ between the position of the sun at solar noon and the equator plane for day n of the year can be approximated by equation (6.5):

$$\delta \approx 23.45° \cdot \sin\left(360° \cdot \frac{284 + n}{365}\right). \tag{6.5}$$

This angular information, combined with the beam irradiance value on a horizontal area gained from meteorological data, leads to the beam irradiance received by a module with a specific tilt angle and orientation on its plane of the array. $E_{\text{beam,POA}}$, is the product of the beam (normal) irradiance, E_{beam}, and the cosine of the beam incidence angle with respect to the surface normal.

The simplest model for the diffuse sky irradiance assumes an isotropic radiance distribution. In this case, the diffuse irradiance into the plane of the array only depends on the module tilt angle and the global horizontal diffuse irradiance (first term in equation (6.6)). The simplest model for ground-reflected irradiance assumes a flat, homogeneously, and isotropically reflecting plane. The contribution to the POA irradiance can then be expressed by the second term of equation (6.6):

$$E_{\text{diff, POA}} = E_{\text{diff,H}}\left(\frac{1 + \cos\beta}{2}\right) + E_{\text{GH}}\rho_{\text{ground}}\left(\frac{1 - \cos\beta}{2}\right). \tag{6.6}$$

Models with even less data requirements only use monthly mean values for **daily global horizontal irradiation** $H_{\text{glob,hor,day}}$ and either the **clearness index** $K_{t,\text{day}}$ or the **daily diffuse horizontal irradiation** $H_{\text{diff,hor,day}}$. These two numbers for each month are available for many locations and can be expanded into hourly beam irradiation and diffuse irradiation values. It is also possible to generate hourly irradiation data for a clear day without any meteorological data at all, only by using the extraterrestrial irradiance together with models for beam irradiance attenuation in a clear atmosphere and corresponding diffuse irradiance models [91].

The extraterrestrial irradiance at the mean earth distance from the sun is termed the **solar constant** E_{SC} (or G_{SC}) with a measured value of 1353 ± 20 W/m². Due to the orbital eccentricity of the earth, the actual extraterrestrial irradiance E_0 varies by about ± 3.3 % over the year (equation (6.7));

$$E_0 = E_{SC}\left(1 + 0.033 \cdot \cos\left(360° \cdot \frac{n}{365}\right)\right), \tag{6.7}$$

where
E_0 extraterrestrial irradiance at sun-earth-distance of day n,
E_{SC} extraterrestrial irradiance at mean sun-earth-distance,
n   day of the year.

Solar radiation is partly absorbed and partly scattered on its way through the atmosphere, which changes its spectrum and its angular distribution. An important concept for beam irradiance attenuation due to inclined incidence is the **air mass** m_{air} (or AM) model (equation (6.8), [92]). The air mass accounts for attenuation due to the atmospheric path length, referenced to an orthogonal beam path ($m_{air} = 1$):

$$m_{air} = \frac{\exp(-0.0001184 \cdot h)}{\sin \alpha_S^* + 0.5057 \cdot (6.080 + \alpha_S^*)^{-1.6364}}, \tag{6.8}$$

where

m_{air} air mass,
h altitude of location, divided by 1 meter,
α_S^* apparent solar altitude angle divided by 1 degree (including atmospheric refraction, max. difference to solar altitude angle is 0.5°).

Equation (6.9) gives an expression [93] for the beam irradiance in the course of the day (upper part), which needs to be scaled according to the given daily horizontal beam irradiation sum (lower part). It can be used to estimate hourly beam irradiation values:

$$\begin{aligned} E_{beam} &\propto E_0 \cdot \exp\left(-\frac{\left(10 - 0.11 \cdot \frac{\phi}{1°}\right) \cdot m_{air}}{0.9 \cdot m_{air} + 9.4}\right), \\ H_{beam,hor,day} &= \int_{sunrise}^{sunset} E_{beam} \cdot \sin \alpha_S \, dt, \end{aligned} \tag{6.9}$$

where

E_{beam} (normal) beam irradiance [W/m2],
E_0 extraterrestrial irradiance on day n [W/m2],
ϕ geographical latitude of the location divided by 1 degree ($-90° \leq \phi \leq 90°$, north is positive),
α_S solar altitude angle ($0° \leq \alpha_S \leq 90°$, 90° is the zenith).

Equation (6.10) gives an expression for the hourly diffuse horizontal irradiation $H_{diff,hor,hour}$ over the day, if the daily diffuse horizontal irradiation $H_{diff,hor,day}$ is known [91]:

$$H_{diff,hor,hour} = H_{diff,hor,day} \cdot \frac{\pi}{24} \cdot \frac{\cos \omega - \cos \omega_S}{\sin \omega_S - \pi \frac{\omega_S}{180°} \cos \omega_S}, \tag{6.10}$$

where

ω hour angle between the local meridian and the solar noon meridian which moves by 15°/hour ($-180° \leq \omega \leq 180°$, afternoon, Latin "post meridiem", p.m. is positive),
ω_S sunset hour angle.

The **sunset hour angle** ω_S (equation (6.11)) can be derived from the upper equation (6.3) by setting the solar altitude angle to zero:

$$\cos \omega_S = -\tan \phi \tan \delta. \tag{6.11}$$

Since the performance of PV modules is nonlinear, all models that only rely on averaged irradiance data over a day or even a month only deliver qualitative yield and PR information.

To illustrate performance losses in realistic cases, exemplary simulations using long term weather data for two different locations have been performed (Figure 6.1). For an intermediate climate with an annual global horizontal irradiation of 4 GJ/m^2 (1100 kWh/m^2), calculated PR losses due to temperature only amount to 1.6 %, while frequent situations with low irradiance lead to associated losses of 1.7 %. The total module performance loss is 7.1 %. In a warm climate with an irradiation of 7.2 GJ/m^2 (2000 kWh/m^2), total performance losses are considerably higher (10.4 %) and dominated by temperature effects. The frequently high irradiance level reduces the losses due to poor low light behavior of the module. On the power plant level, the warm location may also suffer from substantial losses due to soiling, but this is not considered in PR$_{mod}$.

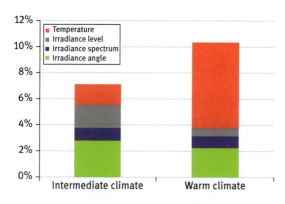

Fig. 6.1: Calculated module performance losses for two different climates.

6.2 Temperature effects

Due to the significant power temperature coefficient of PV modules (Section 2.5), the operating temperature has a strong impact on their performance.

The main parameters that influence module temperature are:
- module absorptance,
- extracted electric power,
- ambient temperature,
- effective sky temperature,
- effective wind speed.

The difference between the absorbed solar irradiance, mainly on the front side of the module, and the extracted electric power leads to the heat generation rate. The equilibrium module temperature additionally depends on the convective and radiative heat exchange with the environment.

Highly efficient solar cells increase the fraction of electric power to absorbed irradiance, thus reducing the heat generation in the module. If inactive module areas on the front and also on the rear side display high reflectance, e.g. in case of a white backsheet, they minimize absorbtance and reduce operation temperature. An AR coating on the glass will increase the absorbance and the power output of the module, and it will slightly increase the operation temperature.

Even low levels of solar irradiance will usually heat the module above ambient air temperature. The effective sky temperature controls the radiant heat transfer between the sky and the module. Wind introduces forced convection, which reduces the temperature difference between the module and ambient air. Free-standing modules, e.g. in ground-mounted power plants, allow back-ventilation and show lower operating temperatures than roof or façade integrated modules.

A variety of models have been proposed for calculating PV module temperatures [94].

Dynamic, multilayer models account for the heat transfer between cells, module surfaces, and environment in detail. They can reproduce the interaction of quickly changing irradiance with the module thermal mass. Static or steady-state single layer models assume permanent thermal equilibrium. Equation (6.12) gives a static model with two fit parameters that capture convective heat transfer [95]:

$$T_{mod} = T_{amb} + \frac{E_{POA}}{c_1} \exp(a + b \cdot v_{wind}), \tag{6.12}$$

with

T_{mod} back surface module temperature [°C],
T_{amb} ambient air temperature [°C],
E_{POA} irradiance into plane of the array (module) [W/m²],
v_{wind} wind velocity measured at 10 m height [m/s],
a,b   parameters for free and forced convective heat transfer,
c_1 constant for dimension adjustment ($c_1 = 1$ W/(m² · °C)).

For a glass/backsheet module in an open rack mount, parameter values are suggested as follows: a = −3.56, b = −0.075 s/m. The solar cell temperature, which is relevant for electric power calculation, is up to 2–3 °C higher at high irradiance levels than the module rear-side temperature measured on the backsheet for an open-rack mount. Center cells of a module have been reported to be about 2 °C warmer than corner cells [96]. A similar model approach reported in [96] is given by equation (6.13) for open-rack mounting. In this model, combined fit parameter values are reported at 25 for a and 6.84 s/m for b:

$$T_{mod} = T_{amb} + \frac{E_{POA}}{c_1} \frac{1}{a + b \cdot v_{wind}}. \tag{6.13}$$

Manufacturers report the nominal operating cell temperature (NOCT) for their modules, which should give an indication for the assessment of power losses in operation conditions. According to IEC 61215 Ed. 2 [57], NOCT applies for open circuit operation at 800 W/m² irradiance, an ambient air temperature of 20 °C, and a wind velocity of 1 m/s in an open rack mounting. NOCT has to be measured within a very narrow range of ambient conditions. The large spread of NOCT values reported in different module data sheets of about 10 °C mainly originates from poor control of measurement conditions; these differences could not be reproduced in a side-by-side monitoring of different modules [97]. The monitoring over 1.5 years revealed that average NOCT values lie in a limited range of 47–48 °C for different modules, while values derived from individual measurements under different conditions show a variation in the range of 7 °C.

The draft edition 3 of the IEC 61215 standard replaces NOCT by the nominal module operating temperature (NMOT) which can be derived from measurements at varying wind speeds, wind directions and ambient temperatures.

In sum, the module performance ratio is positively affected by low cell temperature coefficients (for moderate and warm climates), high cell efficiencies, highly reflective inactive module surfaces, low ambient temperatures, high average wind speed, and an installation that facilitates convection on both front and rear module surfaces.

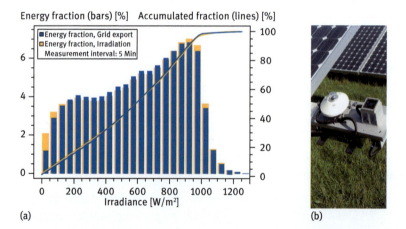

Fig. 6.2: Histogram of POA irradiance for a location in central Europe vs electricity production, measured values (left), pyranometer, and crystalline silicon sensor (right) for irradiance measurement in the POA (Fraunhofer ISE).

6.3 Irradiance level

Next to temperature, module efficiency varies with irradiance. Irradiance values noticeably above 1000 W/m² are only achieved in rare conditions as a combination of sunny skies and bright clouds (Figure 6.2). With increasing irradiance, cell voltage increases (Section 2.2.2), but series resistance losses in cells and strings also rise. In locations with modest global horizontal irradiation, low irradiance levels in the range of several hundred W/m² may frequently occur. At low irradiance, the low light behavior of the cells becomes important (Section 2.6).

Cells and modules are usually designed to maximize nameplate STC module power (and efficiency) at 1000 W/m², and they react quite differently to irradiance deviations from STC. It is therefore important to consider their energy rating for performance and yield predictions at a given location.

6.4 Irradiance angle

6.4.1 Incidence angle modifiers

The short-circuit current of a PV module under beam irradiance declines with increasing incidence angles as compared to normal incidence on the panel surface. The main reason is the **cosine effect**: for a given beam irradiance, the effective collection area of the module declines with the cosine of the angle. On top of the cosine effect, additional losses occur, particularly at incidence angles beyond 60°. For the purpose of module characterization and performance simulation, **incidence angle modifiers** (IAM) are introduced [98, 99]. They provide correction factors for the device short circuit current I_{sc} for incidence angles β from 0° to 90° with respect to normal incidence of 0° (equation (6.14)). The symbol used for the IAM is $K_{\tau\alpha}$, since the modifiers are usually dominated by transmittance (τ) and absorptance (α) effects:

$$K_{\tau\alpha}(\beta) = \frac{I_{sc}(\beta)}{I_{sc}(0°) \cdot \cos(\beta)}. \tag{6.14}$$

An empirical model for the IAM of a given module can be obtained through a polynomial or other functional fit with parameters determined e.g. by the least square method from measured data points [98]. This approach does not require any knowledge on the used materials or coatings:

$$K_{\tau\alpha}(\beta) = \sum_{i=0}^{5} b_i \cdot \beta^i. \tag{6.15}$$

Knowing that the losses are mainly due to increasing reflections at the air-glass interface (Figure 3.29), and to increasing absorption in the cover materials (glass, front-side encapsulant), a theoretical model with physically meaningful parameters can be defined (equation (6.16)):

$$K_{T\alpha}(\beta_0) = \frac{I_{sc}(\beta_0)}{I_{sc}(0°) \cdot \cos(\beta_0)} \approx \frac{T_{0/1}(\beta_0) \cdot T_1(\beta_1) \cdot T_2(\beta_2)}{T_{0/1}(0°) \cdot T_1(0°) \cdot T_2(0°)}, \quad (6.16)$$

where

$K_{T\alpha}$ incidence angle modifier,
β_0 incidence angle from air on module plane,
$I_{sc}(\beta_0)$ short circuit current of module for incidence angle β_0,
$T_{0/1}(\beta_0)$ interface transmittance of air/glass interface for incidence angle β_0,
$T_1(\beta_1)$ bulk transmittance of glass for ray angle β_1,
$T_2(\beta_2)$ bulk transmittance of encapsulant for ray angle β_2.

The ray angle inside the cover material j, which may be glass (j = 1) or encapsulant (j = 2), is given by **Snell's law** (equation (6.17)):

$$\beta_j = \arcsin\left(\frac{\sin(\beta_0)}{n_j}\right). \quad (6.17)$$

Reflection losses at the glass/encapsulant interface are not considered, being relatively small. Additional optical effects as described in Section 5.5 may change with increasing incidence angle, e.g. the reflectance of the cell texture or of a saw-tooth profiled ribbon (Figure 3.6). For this reason, the rights side of equation (6.16) is only an approximation.

For unpolarized beam irradiance incident on a flat interface from material 0 (air) towards material 1 (glass), the interface **transmittance** $T_{0/1}$ is approximated by the **Fresnel equations** for nonabsorbing media (equation (6.18)):

$$T_{0/1}(\beta_0) = 1 - \frac{1}{2}\left[\left(\frac{n_0 \cos(\beta_0) - n_1 \cos(\beta_1)}{n_0 \cos(\beta_0) + n_1 \cos(\beta_1)}\right)^2 + \left(\frac{n_1 \cos(\beta_0) - n_0 \cos(\beta_1)}{n_1 \cos(\beta_0) + n_0 \cos(\beta_1)}\right)^2\right], \quad (6.18)$$

where

$T_{0/1}$ interface transmittance from medium 0 (air) to medium 1 (glass),
n_0 index of refraction of air ($n_0 = 1$),
n_1 index of refraction of glass,
β_0, β_1 ray angles in air and in glass.

An additional, yet smaller loss effect originates from the increase in path length for nonorthogonal rays that pass through the transparent covers. Effective absorption in the bulk glass and encapsulant increases for inclined ray paths (equation (6.19), derived from equation (5.16)). Yet, the path length inside the materials is limited by refraction: even rays that impinge on the air/glass interface at the theoretical limit angle of 90° are refracted towards the normal when they enter the optically denser glass, which limits the maximum path length increase to about one third with respect to the total layer thickness:

$$T_j(\beta_j) = \frac{E_j(d_j)}{E_j(0)} = \exp(-\alpha_j \cdot d_j) = \exp\left(-\alpha_j \cdot \frac{d_{0,j}}{\cos(\beta_j)}\right), \qquad (6.19)$$

where
- β_j ray angle with respect to the module normal inside the cover material j,
- $T_j(\beta_j)$ bulk transmittance of cover material j for direction β_j,
- $E_j(0)$ irradiance inside cover material j immediately after the entry point [W/m²],
- $E_j(d_j)$ irradiance inside cover material j after a path length d_j [W/m²],
- α_j absorption coefficient of cover material j [1/cm],
- d_j path length inside material j [cm],
- $d_{0,j}$ thickness of cover layer j [cm].

Figure 6.3 shows calculated IAM for an air/glass interface and about 1 % **absorptance** in the entire bulk material (glass and encapsulant) according to equation (6.16). IAM data derived from angular short circuit current measurements on four different modules is also displayed. While module 1, 2, and 3 were manufactured with standard glass, module 4 features structured glass. For module 4, the IAM calculated from equation (6.16) do not apply.

For nonlaminated modules (Section 3.2.4), the IAM is governed by the angular transmittance of the front glass including two air/glass interfaces and the angular response of the solar cell.

If the glass bears an AR coating, $T_{0/1}$ is smaller than expressed in equation (6.18) and will usually be determined by measurements. If the glass has an AR texture, equation (6.18) can also not be applied, and interface transmittance is measured or simulated by ray-tracing calculation.

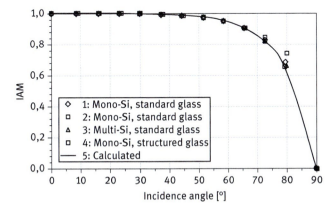

Fig. 6.3: Measured IAM for different modules and calculated IAM for an air/glass interface and 1 % absorptance in the entire bulk material (Fraunhofer ISE).

The IAM from equation (6.14) is strictly speaking a wavelength-dependent quantity, since refractive indeces n_j and absorption coefficients α_j depend on the wavelength. Therefore, equation (6.16) should first be evaluated at different wavelengths, and then the effective IAM should be calculated by spectral integration.

The determination of IAM by outdoor and indoor characterization methods will be defined in the standard IEC 61853-2 currently being drafted [100].

6.4.2 Angular distribution of radiation

For stationary mounted modules, the incidence angle of the direct irradiance changes over the day and the months. Module trackers are most effective at locations with large fractions of direct irradiance. In other locations, for instance in central Europe, the yearly sum of diffuse irradiance may exceed the sum of direct irradiance. The diffuse irradiance from the sky and the ambient displays a complex distribution with high values for the circumsolar sky region, for bright clouds, for the horizon or environmental sections with large albedo (reflectance).

Due to the angle-dependent IAM, module efficiency depends on the current angular distribution of the irradiance with respect to the module plane. For the diffuse part, reflectance losses will generally be higher than for orthogonal incidence, since the effective incidence angle can be estimated in the range of 60°. For direct irradiance, reflectance losses will also be higher than for orthogonal incidence most of the time, if stationary modules are considered.

Figure 6.4 shows calculated angular distribution of the yearly irradiation sum with respect to the module normal for south-oriented modules mounted in the northern hemisphere for different tilt angles and locations. The calculations are based on the models presented in Section 6.1.

Towards orthogonal incidence, the irradiation sum becomes very small due to the limited solid angle under consideration. At the opposite end of the curve (90°), the cosine effect and the sharply increasing air-glass interface reflectance drive the irradiance sum towards zero.

At the location of Freiburg (48° northern latitude) the optimal case with a module tilt angle of 30° with respect to the horizontal is compared with a vertical mounting, e.g. on a building facade. The irradiation sum for vertical mounting integrated over the entire angle range is 30 % lower than for the 30° tilt angle and the mean incidence angle is higher, which leads to additional yield losses. For the location of Cairo, the differences in yield and mean incidence angle are even more pronounced.

Calculated performance losses due to irradiance angle effects have been reported at 3–4 % for stationary modules in optimal orientation [99]. For nonoptimal orientation, e.g. vertically mounted BIPV modules at medium to low latitudes, the performance losses from IAM may reach 5–8 %, depending on orientation and inclination.

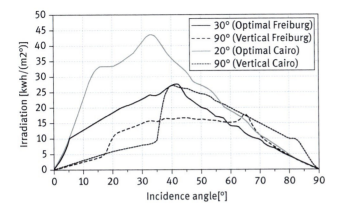

Fig. 6.4: Calculated angular distribution of the yearly sum of the global irradiance in the POA for different module orientations and inclinations for the locations Freiburg (48° northern latitude) and Cairo (30° north. lat.).

For tracked modules in locations with high levels of direct normal irradiation (DNI), the IAM losses are less relevant.

6.5 Irradiance spectrum

The spectral response of the solar cell (Section 2.4) in conjunction with the spectral transmittance of the cover and encapsulation materials (Section 5.5) determines the resulting spectral response of the module.

The actual irradiance spectrum depends on the solar elevation angle with the corresponding air mass and the sky condition. Low solar elevation angles shift the irradiance spectrum towards longer wavelengths, while the absence of direct irradiance leads to a contrary effect. The current irradiance spectrum incident on a module deviates from the AM 1.5 spectrum used for STC characterization and may thus lead to performance losses or gains. These changes can be expressed in a **spectral mismatch factor** as defined in the standard IEC60904-7 [101].

The average mismatch factor defined over a full year thus depends on the particular location and PV technology. If measured spectral irradiance values are available over a full year, the average mismatch factor can be calculated. For crystalline silicon modules, measurements performed for Freiburg, Germany indicate a mismatch factor in the range of +1.1 to +1.4 %, while various thin film PV technologies display values in the range of +0.6 to +3.4 % [102].

6.6 Bifacial irradiance

PV modules that use solar cells with a (semi-)transparent rear side combined with a (semi-)transparent rear module cover show current gains from rear side irradiance. A **bifacial power gain factor** f_{bif} can be used to characterize the power increase stemming from bifacial operation (equation(6.20)). As reference, the same module is operated without rear side irradiance (Section 4.1). f_{bif} can be measured under conditions similar to STC in terms of temperature and spectrum, but with an additional rear side irradiance (e. g. 200 W/m²):

$$f_{bif} = \frac{P_{mod,bif}}{P_{mod,front}}, \qquad (6.20)$$

where

f_{bif} bifacial power gain factor at a specific operating point,
$P_{mod,bif}$ module power in bifacial operation [W],
$P_{mod,front}$ reference module power in monofacial operation [W].

Using the rear-to-front efficiency ratio of single side efficiencies (Section 4.1) at STC, f_η, we introduce a front side efficiency in bifacial operation $\eta_{front,bif}$ (equation (6.21)):

$$P_{mod,bif} = \eta_{front,bif} \left(E_{front} + f_\eta E_{rear} \right), \qquad (6.21)$$

where

f_η rear-to-front efficiency ratio at STC for single side irradiance,
$\eta_{front,bif}$ front side efficiency in bifacial operation.

Equation (6.20) can thus be rewritten to equation (6.22):

$$f_{bif} = \frac{\eta_{front,bif} \left(E_{front} + f_\eta E_{rear} \right)}{\eta_{front} E_{front}}. \qquad (6.22)$$

If we assume that the front side efficiency in bifacial operation as defined by equation (6.21) is similar to the value in monofacial operation, f_{bif} is approximated by equation (6.23). In reality, several effects mentioned in Section 4.2 lead to an efficiency change as soon as additional rear-side irradiance comes into play. In field operation, a higher rear-side parasitic absorptance is expected in bifacial modules, as compared to common monofacial modules with a white backsheet. This will lead to slightly increased operating temperatures.

$$f_{bif} \approx 1 + f_\eta \frac{E_{rear}}{E_{front}}. \qquad (6.23)$$

The rear-side irradiance, E_{rear}, mainly originates from ground-reflected irradiance, E_{ground}, and from diffuse sky irradiance E_{diff}. Depending on the installation, contributions from beam irradiance E_{beam} may occur. The rear-side irradiance is not only influenced by the tilt angle and row distance, but also by the mounting distance to the ground and shading from mounting components. Common environments for ground-mounted PV power plants like grass, bare ground, or sand show reflectance values in the range of 0.15–0.25; a value of 0.2 is usually assumed if no specific information is available. Clean cement, white foils, or snow can provide reflectance values from 0.5 up to 0.9 [103]. If the rear-side irradiance is dominated by ground-reflected irradiance, a linear dependency on the albedo is to be expected [104]. In the field, it is difficult to achieve a homogeneous irradiance on the rear side of a module, due to mounting fixtures and to different view angles for surrounding surfaces, depending on the location within the module area.

For a precise simulation of rear-side irradiance, ray-tracing tools (Fig. 6.5) are required which consider the solar coordinates, the sky condition, and the geometry of the mounting elements. Simulations show that a higher mounting position with respect to the ground can improve the yield gain from bifacial operation. For modules mounted above inclined roofs with a small air gap, bifaciality has little impact on yield, because the space behind the module is heavily shaded. The ground reflectance also has a strong impact, especially at lower module tilt angles.

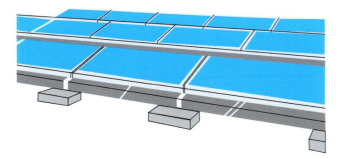

Fig. 6.5: Image of a PV system rendered with the backward raytracer software RADIANCE; the same tool is used to calculate front and rear side irradiance (Fraunhofer ISE).

Bifacial yield gains are usually reported with respect to a similar reference setup which blocks rear side irradiance. When bifacial yield gains are reported, the row setup has to be carefully taken into account. Stand-alone modules can receive much more rear-side irradiance than modules mounted in consecutive rows and may show gains in the range of 25–30 %. The bifacial yield gains reported from a validated simulation of a multirow system on a flat roof with albedo values from 0.2 to 0.6 and f_η = 0.8 vary between 6 to 16 % [105]. Systems performance ratios usually limited to 80–90 % for monofacial modules may thus be extended towards and even beyond 100 % with bifacial modules.

6.7 Energy payback time

The manufacturing of PV modules, starting from the production of metallurgical silicon, consumes a considerable amount of energy. Additional energy is required for the construction of PV power plants. This energy should be returned as quickly as possible during power plant operation. The time period required for energy recovery is the energy payback time (EPBT). When EPBT is specified, the location, especially the annual irradiation needs to be stated.

Due to tremendous progress in material and energy saving along the entire PV production chain and also to conversion efficiency improvements, EPBT could be reduced to values of two to three years even for regions with modest irradiation (Figure 6.6). Mono-Si cells show more than twice the EPBT of poly-Si cells. They require much more energy in production for their higher purity feedstock and more complex crystallization in relation to the achieved efficiency gain effective in the power plant. For the module and system section, the EPBTs are quite similar for both cell technologies, with slight gains on the more efficient mono-Si side.

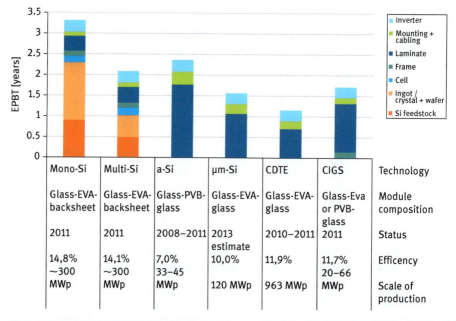

Fig. 6.6: : Calculated energy payback times for different module designs for an annual irradiation of 1000 W/m². Image from [106].

7 References

[1] D. M. Chapin, C. S. Fuller, and G. L. Pearson, A New Silicon p-n Junction Photocell for Converting Solar Radiation into Electrical Power, *J. Applied Physics* 25(5) (1954), 676–677.

[2] J. Perlin, Let It Shine: The 6,000-Year Story of Solar Energy, rev. ed., New World Library, 2013.

[3] M. A. Green, Silicon Photovoltaic Modules: A Brief History of the First 50 Years, *Prog. Photovolt: Res. Appl.* 13 (2005), 447–455.

[4] Photovoltaics Report, Fraunhofer Institute for Solar Energy Systems ISE, September 2015

[5] P. Dupeyrat, C. Menezo, H. Wirth, and M. Rommel, Improvement of PV module optical properties for PV-thermal hybrid collector application, *Solar Energy Materials & Solar Cells* 95 (2011), 2028–2036.

[6] M.A. Green, K. Emery, Y. Hishikawa, W. Warta, and E. D. Dunlop, Solar cell efficiency tables (version 45), *Prog. Photovolt: Res. Appl.* 23 (2015), 1–9.

[7] IEC 60904-3, Photovoltaic devices – Part 3: Measurement principles for terrestrial photovoltaic (PV) solar devices with reference spectral irradiance data. International Electrotechnical Commission (IEC), Geneva, Switzerland 2008; Ed. 2.0.

[8] G. E. Bunea, K. E. Wilson, Y. Meydbray, M. P. Campbell, and D. M. Ceuster, Low light performance of mono-crystalline silicon solar cells, Conference Record of the IEEE 4th World Conference on Photovoltaic Energy Conversion, 2006.

[9] P. Singh and N. M. Ravindra, Temperature dependence of solar cell performance – an analysis, *Solar Energy Materials & Solar Cells* 101 (2012), 36–45.

[10] K. Mertens, *Photovoltaics: Fundamentals, Technology and Practice*, Wiley, 2014.

[11] J. W. Bishop, Computer simulation of the effects of electrical mismatches in photovoltaic cell interconnection circuits, *Solar Cells* 25(1) (1988), 73–89.

[12] F. Fertig, S. Rein, M. C. Schubert, and W. Warta, Impact of Junction Breakdown in Multi-Crystalline Silicon Solar Cells on Hot Spot Formation and Module Performance, 26th European Photovoltaic Solar Energy Conference and Exhibition 2011, pp. 1168–1178.

[13] A. Luque (ed.) and S. Hegedus (co-ed.), *Handbook of Photovoltaic Science and Engineering*, 2nd ed. Wiley, 2010.

[14] M. A. Green, *High Efficiency Silicon Solar Cells*, Trans. Tech. Publications, 1987, p. 99.

[15] W. Shockley and H. J. Queisser, Detailed Balance Limit of Efficiency of p-n Junction Solar Cells, *J. Applied Physics* 32 (1961), 510–519.

[16] M. A. Green, Silicon wafer-based tandem cells: The ultimate photovoltaic solution? Proc. SPIE 8981, Physics, Simulation, and Photonic Engineering of Photovoltaic Devices III 2014.

[17] N. H. Reich, W. G. J. H. M. van Sark, E. A. Alsema, R. W. Lof, R. E. I. Schropp, W. C. Sinke, and W. C. Turkenburg, Crystalline silicon cell performance at low light intensities. *Solar Energy Materials and Solar Cells* 93(9) (2009), 1471–1481.

[18] P. Grunow, S. Lust, D. Sauter, V. Hoffmann, C. Beneking, B. Litzenburger, and L. Podlowski, Weak light performance and annual yields of PV modules and systems as a result of the basic parameter set of industrial solar cells. 19th European Photovoltaic Solar Energy Conference, Paris, 2004.

[19] V. A. Popovich, A. Yunus, M. Janssen, I. M. Richardson, and I. J. Bennett, Effect of silicon solar cell processing parameters and crystallinity on mechanical strength, *Solar Energy Materials and Solar Cells* 95 (2011), 97–100.

[20] G. Coletti, N. J. C. M. van der Borg, S. De Iuliis, C. J. J. Tool, and L. J. Geerligs, Mechanical strength of silicon wafers depending on wafer thickness and surface treatment. 21st European Photovoltaic Solar Energy Conference and Exhibition, Dresden, 2006.

[21] E. van Kerschaver and G. Beaucarne, Back-contact Solar Cells: A Review. *Progress in Photovoltaics: Research and Applications* 14(2) (2005), 107–123.

[22] H. von Campe, S. Huber, S. Meyer, S. Reiff, and J. Vietor, Direct Tin-Coating of the Aluminum Rear Contact by Ultrasonic Soldering, 27th European Photovoltaic Solar Energy Conference, Frankfurt, 2012.

[23] M. A. Green and M. J. Keevers, Optical properties of intrinsic silicon at 300 K. *Progress in photovoltaics: research and applications* 3 (1995), 189–192.

[24] K. R. McIntosh and S. C. Baker-Finch, OPAL 2: Rapid Optical Simulation of Silicon Solar Cells, 38th IEEE Photovoltaic Specialists Conference (PVSC), 2012.

[25] J. Nievendick, M. Stocker, J. Specht, W. Glover, M. Zimmer, and J. Rentsch, Application of the inkjet-honeycomb-texture in silicon solar cell production, *Energy Procedia* 27 (2012), 385–389.

[26] Grohe A. (2008). "Einsatz von Laserverfahren zur Prozessierung von kristallinen Silicium-Solarzellen". Dissertation, Universität Konstanz, Fakultät für Physik, Germany.

[27] N. Tucher, A. Volk, J. Seiffe, H. Hauser, C. Müller, und B. Bläsi, Large-area honeycomb texturing of si-solar cells via nanoimprint lithography. 29th European Photovoltaic Solar Energy Conference and Exhibition 2014, pp. 1006–1011.

[28] H. Hauser, N. Tucher, K. Tokai, P. Schneider, C. Wellens, et al., Development of nanoimprint processes for photovoltaic applications, *J. Micro/Nanolith. MEMS MOEMS.* 14(3) (2015), 031210 1–6.

[29] J. Ziegler, J. Haschke, T. Käsebier, L. Korte, and A. N. Sprafke, Wehrspohn RB. Influence of black silicon surfaces on the performance of back-contacted back silicon heterojunction solar cells, *Optics Express* 22(S6) (2014), A1469–A1476.

[30] C. Peike, Thermal Influence on the Photochemical Aging Behaviour of Ethylene-Based PV Module Encapsulants, 29th European Photovoltaic Solar Energy Conference and Exhibition, Amsterdam, 2014.

[31] R. Preu, G. Kleiss, K. Reiche, and K. Bücher, PV-Module reflection losses: measurement, simulation and influence on energy yield and performance ratio, 13th European Photovoltaic Solar Energy Conference and Exhibition, Nice, 1995.

[32] I. Chung, H. Y. Son, H. Oh, U. Baek, N. Yoon, W. Lee E. Cho, and I. Moon, Light Capturing Film on interconnect ribbon for current gain of crystalline silicon PV modules, 39th IEEE Photovoltaic Specialists Conference, Tampa, 2013.

[33] M. Ebert, M. Seckel, L. Böttcher, M. Hendrichs, F. Clement, I. Dürr, D. Biro, U. Eitner, and M. Schneider-Ramelow, Robust module integration of back contact cells by interconnection adapters, 29th European Photovoltaic Solar Energy Conference and Exhibition, Amsterdam, 2014.

[34] Y. Yuan and T. R. Lee, Contact Angle and Wetting Properties, in: G. Bracco, B. Holst (eds.), *Surface Science Techniques*, Springer Series in Surface Sciences 51, Springer, Berlin Heidelberg, 2013.

[35] B. A. Smith and L. J. Turbini, Characterizing the Weak Organic Acids Used in Low Solids Fluxes, *J. Electronic Materials* 28(11) (1999), 1299–1306.

[36] J. Wendt, M. Träger, M. Mette, A. Pfennig, and B. Jäckel, The link between mechanical stress induced by soldering and micro damages in silicon solar cells, 24th European Photovoltaic Solar Energy Conference and Exhibition, Hamburg, 2009.

[37] A. M. Gabor, M. Ralli, S. Montminy, L. Alegria, C. Bordonaro, J. Wood, and L. Felton, Soldering Induced Damage to Thin Si Solar Cells and Detection of Cracked Cells in Modules, 21st European Photovoltaic Solar Energy Conference and Exhibition, Dresden, 2006.

[38] D. Eberlein, P. Schmitt, and P. Voos, Metallographic Sample Preparation of Soldered Solar Cells, *Practical Metallography* 48 (2011), 239–260.

[39] S. Kajari-Schröder, I. Kunze, and M. Köntges, Criticality of cracks in PV modules, *Energy Procedia* 27 (2012), 658–663.

[40] A. Schneider, L. Rubin, and G. Rubin, Solar cell efficiency improvement by new metallization techniques – the DAY4™electrode concept. Conference Record of the IEEE 4th World Conference on Photovoltaic Energy Conversion 2006. pp. 1095-1098.

[41] T. Söderström, P. Papet, and J. Ufheil, Smart wire connection technology, 28th European Photovoltaic Solar Energy Conference and Exhibition, Paris, 2013.

[42] J. Glazer, Microstructure and mechanical properties of Pb-free solder alloys for low-cost electronic assembly, a review, *J. Electronic Materials* 23(8), 1994, 693–700.

[43] F. Hua, Z. Mei, and J. Glazer, Eutectic Sn-Bi as an alternative to Pb-free solders, Electronic Components and Technology Conference, 1998.

[44] H. Tanaka, L. f. Qun, O. Munekata, T. Taguchi, and T. Narita, Elastic Properties of Sn-Based Pb-Free Solder Alloys Determined by Ultrasonic Pulse Echo Method, *Materials Transactions* 46(6) (2005), 1271–1273.

[45] M. Abtew and G. Selvaduray, Lead-free Solders in Microelectronics, *Materials Science and Engineering* 27 (2000), 95–141.

[46] D. H. Kim, Reliability Study of SnPb and SnAg Solder Joints in PBGA Packages, Dissertation, Faculty of the Graduate School of The University of Texas at Austin, 2007.

[47] W. J. Plumbridge, C. R. Gagg, and S. Peters, The creep of lead–free solders at elevated temperatures, *J. Electronic Materials* 30(9) (2001), 1178–1183.

[48] N. Hiyoshi, T. Itoh, and M. Sakane, Thermo-Mechanical Fatigue Life Evaluation for Sn-Pb and Sn-Ag Solders, 18th European Microelectronics and Packaging Conference (EMPC), 2011.

[49] E. Suhir, Y. C. Lee, and C. P. Wong (eds.), *Micro- and Opto-Electronic Materials and Structures: Physics, Mechanics, Design, Reliability, Packaging*, Springer 2007.

[50] T. Geipel and U. Eitner, Electrically Conductive Adhesives-An Emerging Interconnection Technology for High-Efficiency Solar Modules, *Photovoltaics International* 21 (2013), 27–33.

[51] U. Eitner and L. Rendler, The Mechanical Theory Behind the Peel Test, *Energy Procedia* 55 (2014), 331–335.

[52] DIN EN 50461, Solar cells – Datasheet information and product data for crystalline silicon solar cells, 2007.

[53] E. Le Bourhis, *Glass: Mechanics and Technology*, Weinheim, Wiley, 2014.

[54] P. Sánchez-Friera, D. Montiel, J. F. Gil, J. A. Montañez, and J. Alonso, Daily power output increase of over 3 % with the use of structured glass in monocrysalline silicon PV modules, IEEE 4th World Conference on Photovoltaic Energy Conversion, Waikoloa, 2006.

[55] C. Ballif, J. Dicker, D. Borchert, and T. Hofmann, Solar glass with industrial porous SiO2 antireflection coating: measurements of photovoltaic module properties improvement and modelling of yearly energy yield gain, *Solar Energy Materials & Solar Cells* 82 (2004), 331–344.

[56] E. Klimm, T. Lorenz, and K.-A.Weiss, Can Anti-Soiling Coating on Solar Glass Influence the Degree of Performance Loss over Time of PV Modules Drastically? 28th European Photovoltaic Solar Energy Conference and Exhibition, 2013.

[57] IEC 61215, Crystalline silicon terrestrial photovoltaic (PV) modules – Design qualification and type approval. International Electrotechnical Commission (IEC), Geneva, Switzerland 2005; Ed. 2.0.

[58] M. D. Kempe, Rheological and Mechanical Considerations for Photovoltaic Encapsulants, DOE Solar Energy Technologies Program Review Meeting, Denver, 2005.

[59] K. Agroui, G. Collins, and J. Farenc, Measurement of glass transition temperature of crosslinked EVA encapsulant by thermal analysis for photovoltaic application, *Renewable Energy* 43 (2012), 218–223.

[60] R. Polansky, P. Prosr, and M. Pinkerova, Investigation of encapsulant material used in photovoltaic modules by thermal analysis, Proceedings of the 22nd International DAAAM Symposium 2011, Volume 22, No. 1.

[61] E. F. Cuddihy, C. D. Coulbert, R. H. Liang, A. Gupta, P. Willis, and B. Baum, Applications of Ethylene Vinyl Acetate as an Encapsulation Material for Terrestrial Photovoltaic, Jet Propulsion Laboratory DOE/JPL/1012–87 1983.

[62] M. Kempe, Overview of scientific issues involved in selection of polymers for PV applications, 37th IEEE Photovoltaic Specialists Conference, Seattle, 2011.

[63] A. Beinert, C. Peike, I. Dürr, M. D. Kempe, G. Reiter, and K.-A. Weiß, The influence of the additive composition on degradation induced changes in poly(ethylene-co-vinyl acetate) during photochemical aging, 29th European Photovoltaic Solar Energy Conference and Exhibition, EU PVSEC 2014.

[64] C. Peike, I. Hädrich, K. A. Weiss, and I. Dürr, Overview of PV module encapsulation materials, *Photovoltaics International* 19 (2013).

[65] R. Einhaus, K. Bamberg, R. de Franclieu, and H. Lauvray, New industrial solar Cell Encapsulation (NICE) technology for PV module fabrication at drastically reduced costs, 19th European Photovoltaic Solar Energy Conference and Exhibition, Paris, 2004.

[66] ASTM D6862 – 11, Standard Test Method for 90 Degree Peel Resistance of Adhesives, ASTM International 2011.

[67] DIN EN 28510-1, Adhesives – Peel test for a flexible-bonded-to-rigid test specimen assembly – Part 1: 90 °peel 2014.

[68] F. J. Pern and S. H. Glick, Adhesion Strength Study of EVA Encapsulants on Glass Substrates, National Center for Photovoltaics, National Renewable Energy Laboratory, Solar Program Review Meeting, Denver, Colorado, 2003.

[69] ASTM D2765 – 11 Standard Test Methods for Determination of Gel Content and Swell Ratio of Crosslinked Ethylene Plastic, 2011.

[70] C. Hirschl, M. Biebl–Rydlo, M. DeBiasio, W. Mühleisen, L. Neumaier, W. Scherf, G. Oreski, G. Eder, B. Chernev, W. Schwab, and M. Kraft, Determining the degree of crosslinking of ethylene vinyl acetate photovoltaic module encapsulants – A comparative study, *Solar Energy Materials and Solar Cells* 116 (2013), 203–218.

[71] H. Y. Li, L. E. Perret-Aebi, R. Théron, C. Ballif, and Y. Luo, Lange RFM. Optical transmission as a fast and non-destructive tool for determination of ethylene-co-vinyl acetate curing state in photovoltaic modules, *Prog. Photovolt: Res. Appl.* 21 (2013), 187–194.

[72] R. F. M. Lange, Y. Luo, R. Polo, and J. Zahnd J., The lamination of (multi)crystalline and thin film based photovoltaic modules, *Prog. Photovolt: Res. Appl.* 19 (2011), 127–133.

[73] EN 50380, Datasheet and nameplate information for photovoltaic modules, European Standard 2003.

[74] International Technology Roadmap for Photovoltaic (ITRPV). 6th ed., 2015.

[75] IEC 60904-9, Photovoltaic devices – Part 9: Solar simulator performance requirements. International Electrotechnical Commission (IEC), Geneva, Switzerland 2007; Ed. 2.0.

[76] D. Polverini, G. Tzamalis, and H. Müllejans, A validation study of photovoltaic module series resistance determination under various operating conditions according to IEC 60891, *Prog. Photovolt: Res. Appl.* 20 (2011), 650–660.

[77] D. Sera and R. Teodorescu, Robust series resistance estimation for diagnostics of photovoltaic modules, Industrial Electronics, 35th Annual Conference of IEEE , Porto, 2009.

[78] IEC 60891, Photovoltaic devices – Procedures for temperature and irradiance corrections to measured I-V characteristics. International Electrotechnical Commission (IEC) 2009; Ed. 2.0.

[79] V. d'Alessandro, P. Guerriero, S. Daliento, and M. Gargiulo, A straightforward method to extract the shunt resistance of photovoltaic cells from current–voltage characteristics of mounted arrays, *Solid-State Electronics* 63 (2011), 130–136.

[80] IEC 61853-1, Photovoltaic (PV) module performance testing and energy rating – Part 1: Irradiance and temperature performance measurements and power rating. International Electrotechnical Commission (IEC), Geneva, Switzerland 2011; Ed. 1.0.

[81] D. L. King, B. R. Hansen, J. A. Kratochvil, and M. A. Quintana, Dark Current-Voltage Measurements on Photovoltaic Modules as a Diagnostic or Manufacturing Tool. Proc. 26th IEEE Photovoltaic Specialists Conference, USA, 1997.

[82] T. Fuyuki and A. Kitiyanan, Photographic diagnosis of crystalline silicon solar cells utilizing electroluminescence, *Applied Physics A* 96 (2009), 189–196.

[83] T. Potthoff, K. Bothe, U. Eitner, D. Hinken, and M. Köntges, Detection of the voltage distribution in photovoltaic modules by electroluminescence imaging, *Prog. Photovolt: Res. Appl.* 18 (2010), 100–106.

[84] I. Haedrich, U. Eitner, M. Wiese, and H. Wirth, Unified methodology for determining CTM ratios: systematic prediction of module power, *Solar Energy Materials & Solar Cells* 131 (2014), 14–23.

[85] A.G. Aberle, W. Zhang, and B. Hoex (2011), Advanced loss analysis method for silicon wafer solar cells, 1st SiliconPV Conference, Germany.

[86] IEC 61730-1 Photovoltaic (PV) module safety qualification – Part 1: Requirements for construction. International Electrotechnical Commission (IEC), Geneva, Switzerland 2004; Ed. 1.0.

[87] D. C. Miller, M. D. Kempe, C. E. Kennedy, and S. R. Kurtz, Analysis of transmitted optical spectrum enabling accelerated testing of multijunction concentrating photovoltaic designs, *Opt. Eng.* 50(1) (2011), 013003.

[88] Y. S. Khoo, T. Walsh, F. Lu, and A. G. Aberle, Method for quantifying optical parasitic absorptance loss of glass and encapsulant materials of silicon wafer based photovoltaic modules, *Solar Energy Materials & Solar Cells* 102 (2012), 153–159.

[89] K. R. McIntosh, J. N. Cotsell, J. S. Cumpston, A. W. Norris, N. e. Powell, and B. M. Ketola, An optical comparison of silicone and EVA encapsulants for conventional silicon PV modules: A ray-tracing study, 34th IEEE Photovoltaic Specialists Conference, Philadelphia, 2009.

[90] E. Fornies, F. Naranjo, M. Mazo, and F. Ruiz, The influence of mismatch of solar cells on relative power loss of photovoltaic modules, *Solar Energy* 97 (2013), 39–47.

[91] J. A. Duffie and W. A. Beckman, *Solar Engineering of Thermal Processes* 4th ed., John Wiley & Sons, Hoboken, New Jersey, 2013.

[92] F. Kasten and A. T. Young, Revised optical air mass tables and approximation formula, *Applied Optics* 28(22) (1989), 4735–4738.

[93] H. G. Beyer, H. M. Henning, J. Luther, and K. R. Schreitmüller, The monthly average daily time pattern of beam radiation, *Solar Energy* 47(5) (1991), 347–353.

[94] A. Q. Jakhrani, A. K. Othman, A. R. H. Rigit, and S. R. Samo, Comparison of Solar Photovoltaic Module Temperature Models, *World Applied Sciences Journal* 14 (2011).

[95] D. L. King, W. E. Boyson, and J. A. Kratochvill, Photovoltaic Array Performance Model, SANDIA REPORT 2004; 3535.

[96] D. Faiman, Assessing the outdoor operating temperature of photovoltaic modules, *Progress in Photovoltaics* 16(4) (2008), 307–315.

[97] M. Muller, B. Marion, and J. Rodriguez, Evaluating the IEC 61215 Ed. 3 NMOT Procedure Against the Existing NOCT Procedure with PV Modules in a Side-by-Side Configuration. Photovoltaic Specialists Conference (PVSC), 38th IEEE 2012, pp. 697–702.

[98] W. De Soto, S. A. Klein, and W. A. Beckman, Improvement and validation of a model for photovoltaic array performance, *Solar Energy* 80 (2006), 78–88.

[99] N. Martin and J. M. Ruiz, Calculation of the PV modules angular losses under field conditions by means of an analytical model, *Solar Energy Materials & Solar Cells* 70 (2001), 25–38.

[100] FprEN 61853-2 Photovoltaic (PV) module performance testing and energy rating – Part 2: Spectral response, incidence angle and module operating temperature measurements, draft standard from IEC 82/606/CDV 2010.

[101] IEC 60904-7, Photovoltaic devices – Part 7: Computation of the spectral mismatch correction for measurements of photovoltaic devices. International Electrotechnical Commission (IEC) 2008; Ed. 3.0.

[102] D. Dirnberger, G. Blackburn, B. Müller, and C. Reise, On the impact of solar spectral irradiance on the yield of different PV technologies, *Solar Energy Materials and Solar Cells* 132 (2015), 431–442.

[103] V. Quaschning, *Regenerative Energiesysteme Technologie – Berechnung – Simulation*, Carl Hanser Verlag, München, 2013.

[104] U. A. Yusufoglu, T. H. Lee, T. M. Pletzer, A. Halm, L. J. Koduvelikulathu, C. Comparotto, R. Kopecek, and H. Kurz, Simulation of energy production by bifacial modules with revision of ground reflection, *Energy Procedia* 55 (2014), 389–395.

[105] C. Reise and A. Schmid, Realistic Yield Expectations for Bifacial PV Systems – an Assessment of Announced, Predicted and Observed Benefits, 30th European Photovoltaic Solar Energy Conference and Exhibition, Hamburg, 2015.

[106] Data: M. J. de Wild-Scholten 2013, Graph: PSE AG 2014, Photovoltaics Report, Fraunhofer Institute for Solar Energy Systems ISE, 2014

Part II: Crystalline Silicon Module Reliability

8 Characterization of modules and degradation effects

In the following chapter, the characterisation of PV modules shall be described from a quality assurance point of view as well as from a degradation analysis point of view. While this chapter focuses on the specific analytical techniques which are available for PV module analysis, later chapters will give an in-depth description of the loads which impact on PV modules during their lifetime (Chapter 9) as well as of the testing procedures and required equipment for accelerated aging and reliability testing of PV materials and modules (Chapters 10–12). A detailed description of the various degradation effects which can occur in PV modules is beyond the scope of this book and can be found elsewhere [107–111].

8.1 Destructive analytics

In the following, various destructive analytical techniques are presented and explained as well as their application to analyse PV modules and materials.

8.1.1 Gel content analysis

The degree of cross-linking of EVA encapsulated PV modules is usually quantified indirectly by measurement of the residual gel content after solvent extraction [112, 113]. The cross-linking degree is an important quality assurance measure, as it gives information about the of the lamination process. It also gives information about potential peroxide residues in the encapsulation material which may cause problematic degradation reactions. For a detailed description of the gel content measurement, please refer to Section 3.2.5.2 of Chapter 3.

8.1.2 Differential scanning calorimetry DSC

Differential scanning calorimetry (DSC) is a thermoanalytical technique which measures the different amounts of heat required to increase the temperature of a sample and a reference. For PV modules, this method can be used to determine important properties such as the residual cross-linking agent content in the encapsulation material. For a brief description, please refer to Section 3.2.5.2.

8.1.3 Dynamic mechanical analysis DMA

The dynamic mechanical analysis is a technique which measurer the resulting strain in a sample to which a stress is applied at a certain temperature and with a certain frequency. This way, for instance, the glass transition temperature can be measured. For a detailed description, please refer to Section 3.2.3.

8.1.4 Energy-dispersive x-ray spectroscopy EDX

An elemental analysis of samples can be carried out by EDX microanalysis which relies on the characteristic x-ray spectrum generated by a bombardment of the sample with a focused x-ray beam. This beam of electrons induces the liberation of electrons from the inner shells of an atom which are than refilled by electrons from outer shells. This process is schematically shown in Figure 8.1.

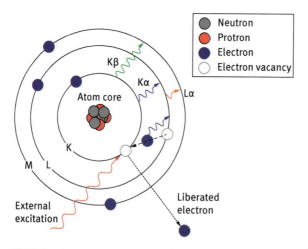

Fig. 8.1: Schematic description of electronic processes during an EDX experiment. Adapted according to [114].

The Rutherford–Bohr atom model describes electrons orbiting the positive nucleus. In the ground state the number of orbital electrons equals the number of protons in the nucleus, given by the atomic number Z. These orbital states have specific energies and are defined by quantum numbers. With increasing Z, orbits are occupied aiming for a minimum energy, those nearest the nucleus, and therefore the most tightly bound, being filled first. Orbital energy is determined mainly by the principal quantum number (n). As shown in Figure 8.2, the shell closest to the nucleus (n = 1) is known as the K shell, followed by the L shell (n = 2) and the M shell (n = 3), etc. [115].

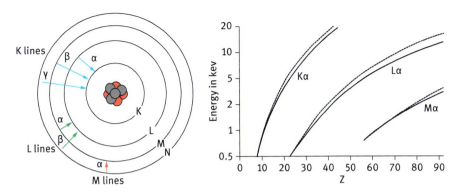

Fig. 8.2: The K, L and M lines are generated via the electronic transition between one α two β or three γ shells (left). Energies of principal characteristic lines (straight lines) and their excitation energies (dotted lines) (right) 123].

8.1.5 Auger electron spectroscopy AUGER

Auger electron spectroscopy is a surface specific technique utilising the emission of low energy electrons in the *Auger process* and is one of the most commonly employed surface analytical techniques for determining the composition of the surface layers of a sample. As shown in Figure 8.3, the electrons which are detected by Auger spec-

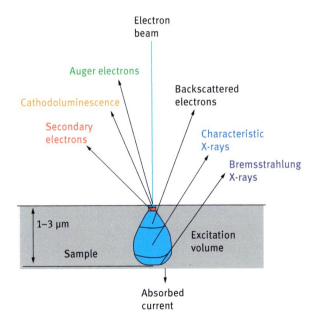

Fig. 8.3: While the Auger electrons are generated from the first monolayers of a sample, cathodoluminescence, secondary electrons, x-rays stem from deeper layers.

troscopy stem from the very top of a sample, typically from the first monolayers. Thus, this method for elemental analysis of surfaces exhibits the highest surface sensitivity of the presented methods and can give a quantitative compositional analysis of the surface region of specimens, by comparison with standard samples of known composition. While the Auger electrons are generated from the first monolayers of a sample, cathodoluminescence, secondary electrons, x-rays stem from deeper layers.

8.1.6 Peel testing

Peel testing is used in photovoltaics to determine the adhesion of interconnector ribbons to solar cell metallization, but also of the adhesion of the other PV module components to each other. For a detailed description of peel tests, please refer to Section 3.2.5.1.

8.2 Non-destructive analytics

In this section, nondestructive analytical techniques are presented which have the advantage of functioning without extensive sample preparation and which allow the modules to stay intact, e.g. for additional testing.

8.2.1 IV-curve measurements

The power output of PV modules is determined by the so-called flasher measurement. This test is performed under standardized test conditions at 25 °C, with 1000 W/m² irradiation and an air mass 1.5 spectrum as by the norm DIN IE 60904-5. For a detailed description of IV measurements please see Section 4.1 of Chapter 4.

8.2.2 Internal and external quantum efficiency IQE and EQE measurements

The external quantum efficiency (EQE) can be defined as the ratio of the number of charge carriers collected by the solar cell and the number of photons of a given energy reaching a solar cell (incident photons). In contrast, the internal quantum efficiency (IQE) is the ratio of the number of charge carriers collected by the solar cell to the number of photons of a given energy that reach a solar cell from outside and are absorbed by the cell.

The measurement of the external and internal quantum efficiency (EQE and IQE) allows for an evaluation of the spectral response of a solar cell. The setup of an EQE measurement system used at the Fraunhofer ISE is shown in Figure 8.4.

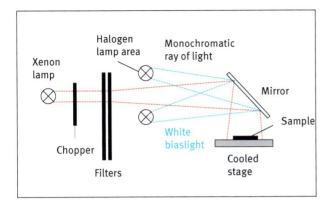

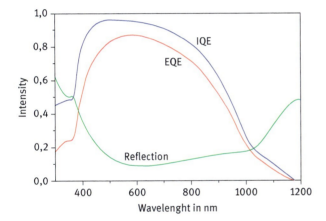

Fig. 8.4: Upper figure, a schematic description of the setup of an EQE measurement system utilizing a xenon lamp as light source is shown. Typical curves of the internal and external quantum efficiency as well as the reflection of a solar cell are given in the lower graph [116].

A xenon lamp can be used as the radiation source. The wavelength selection is carried out using an optical filter and the light modulation is performed by a chopper. Typical curves of the internal and external quantum efficiency and the reflection of a solar cell as shown in Figure 8.4 demonstrate that the IQE and the EQE are connected via

$$\text{IQE} = \text{EQE}\frac{1}{1-R} \qquad (8.1)$$

in which R is the reflection of the solar cell.

8.2.3 Photoluminescence PL imaging

Luminescence imaging of silicon samples utilizes the luminescence emitted from the surface of a sample as a result of radiative recombination of electrons which stem from a precious excitation. In addition to electroluminescence (EL) imaging described above, photoluminescence (PL) imaging can be used for PV module characterization with a focus on the solar cell properties. The difference between these two techniques is the way of luminescence excitation: While PL imaging uses a UV light source EL utilizes voltage. Due to these different excitation processes, the factors influencing the efficiency of electron generation are different. Figure 8.5 displays the set-up of a PL stage consisting of a widened laser beam for the solar cell excitation. The resulting luminescence is detected by a camera with a lock filter that decouples the reflected laser light. PL imaging uses optical excitation, which allows application to a wider range of samples, including bricks, as-cut wafers, and partially processed wafers. The PL intensity is mainly reduced by impaired silicon cell properties such as a deteriorated AR-coating or emitter region

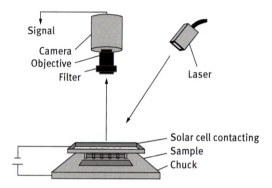

Fig. 8.5: Set up of a PL stage consisting of a widened laser which excites the solar cell. The resulting luminescence is detected by a camera with a lock filter that decouples the laser light which is reflected by the sample [116].

Only in the case of a fully processed solar cell or a module can excitation be carried out by an applied voltage instead of a laser light, allowing for electroluminesce imaging. Due to its simplicity, EL imaging is widely used for module inspection. EL intensity is, among other factors, limited by the conduction properties of the solder bonds and the grid and therefore a corroded or insufficient adhesive metallization to the emitter reduces the EL signal drastically. The combination of both techniques allows the characterization of the cell under different operational conditions as a basis for the evaluation of various cell parameters such as the series resistance R_s or the dark current [116]. Photoluminescence and series resistance of the laminates were measured with a self-constructed test device, described by Haunschild [116].

8.2.4 Gloss- and colour measurement

Yellowness can by defined as the "deviation in chroma from whiteness in the dominant wavelength range from 570 to 580 nm". The Yellowness Index (YI) is the magnitude of yellowness relative to a white standard under illumination by a standard light source [117].

The Yellowness Index is a measure for quantifying the subjective perception of colors by the human eye. This index is most commonly used to evaluate colour changes in a material, e.g. caused by outdoor exposure or accelerated aging.

It can be determined via different standards given by the Commission Internationale de l'Ecleirage (CIE) or by the American Society for Testing and Materials (ASTM). The determination of the Yellowness Index according to the CIE standard will be discussed in following and is visualized in Figure 8.6.

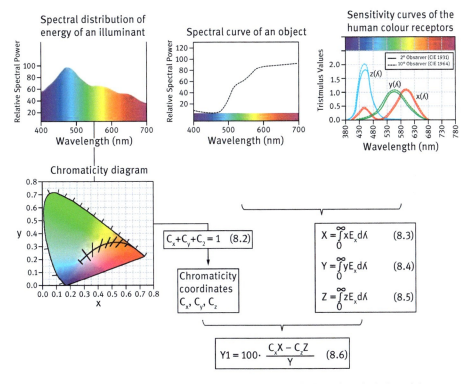

Fig. 8.6: The determination of the yellowness index is accomplished via the calculation of the chromaticity coordinates of the illuminant and the calculation of the tristimulus values taking the spectral curve of the investigated object and the sensitivity curves of the human color receptors into account [118].

Three things are crucial for the perception of a colour:
- the light source,
- the investigated object,
- an observer.

Since this perception of a colour highly depends on the character of the selected light source, illuminants are standardized according to their colour temperature.

Some of the most important standard illuminants are
- A (incandescent, 2856 K),
- C (average daylight, 6774 K),
- D65 (daylight, 6504 K),
- F (fluorescent).

The colour temperature of a specific illuminant is determined by the spectral distribution of its radiation energy. The chromaticity coordinates can be directly obtained via a chromaticity diagram, which is shown in Figure 8.6. Due to the relationship

$$C_X + C_Y + C_Z = 1, \qquad (8.2)$$

only two chromaticity values are needed for an unambiguous assignment, thus chromaticity are usually plotted in a 2D diagram.

Mostly C or D illuminants with a colour temperature above 6000 K are utilized for spectroscopic measurements.

In addition to the character of the illuminant, the spectral curve of the investigated object and the sensitivity curves of the human colour receptors have to be considered.

A specific colour consists of proportions of the variables X, Y and Z, which are called the tristimulus values. The human eye perceives colour via three receptors, which detect the proportions of the light. Those receptors primarily detect the reddish purple ($\bar{x}$), the green ($\bar{y}$) and the blue ($\bar{z}$) components of the light. The sensitivity distribution of the receptors depends on the angle of incidence of the light. For calculations either a 2° (2° observer) or a 10° (10° observer) angle are considered. To quantify the human perception of a light stimulus by an object, the product of these values and the energy distribution E_λ of the sample stimulus, i.e. reflected, transmitted or emitted light, are integrated over a wavelength range as shown in equations (8.3)–(8.5). This way, the tristimulus values X, Y and Z are obtained. The chromaticity coordinates can be directly obtained via a chromaticity diagram, which is shown in Figure 8.6 or calculated with the equations 8.2 to 8.6.

8.2.5 Raman spectroscopy

The Raman effect originates from the interaction between light and matter, resulting in a wavelength shift of the scattered compared to the incident light.

Light interacting with matter may either be absorbed, scattered or transmitted. Absorption only occurs when the energy gap between the molecule's ground and excited state matches the energy of the incident photon. In this case, the molecule is promoted to an excited state, and the loss of energy can be detected to quantify the amount of absorption. With regard to the scattered light, the wavelength of the major part of the scattered light is identical to that of the incident light. But a minor part exhibits a shift of the wavelength which is induced by molecular vibrations and rotations – this phenomenon is referred to as Raman scattering.

A first understanding of the Raman effect can be obtained from energy scheme considerations, which are visualized in Figure 8.7.

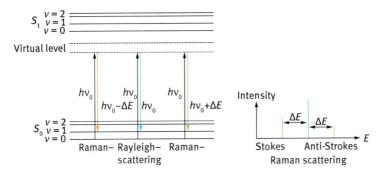

Fig. 8.7: Energy scheme and frequency spectrum showing the Raman and the Rayleigh scattering. Left: electronic transitions between vibration levels of the ground state S_0 and the virtual level, resulting in Raman and Rayleigh scattering. Right: the intensities of the Stokes and the anti-Stokes scattering, which exhibit an energy shift of the same amount with respect to the initial energy, are small compared to the Rayleigh scattering. The intensity of the Stokes scattering is higher than that of the anti-Stokes scattering due to the occurrence fewer molecules in an excited vibrational state [118].

If a molecule is hit by a photon of the energy $E = h\nu_0$ of an energy smaller than $E(S_0 \rightarrow S_1)$, the molecule is excited from the ground state S_0 to a virtual level. This molecule can thereafter either relax into the ground state, emitting a photon with the initial energy $h\nu_0$ (Figure 8.7: blue lines) or into an excited vibrational level under emission of a photon with a reduced energy $h\nu_0 - \Delta E$ (Figure 8.7: orange lines). Here, the first case describes an elastic scattering process – the Rayleigh scattering. The second case describes the inelastic Stokes–Raman scattering. In addition to the Stokes–Raman scattering, a second inelastic scattering process can occur under the prerequisite of molecules in an excited vibrational state: The anti-Stokes–Raman scattering. Anti-Stokes–Raman scattering occurs when initially excited molecules are further excited into a virtual level and relax into the ground state emitting a photon of the energy $h\nu_0 + \Delta E$. As shown in Figure 8.7 (right), the intensity of the Stokes scattering is under ambient conditions much higher than that of the anti-Stokes scattering due to the small amount of molecules in an excited vibrational state. At room temperature, most molecules are in the lowest vibrational energy level. But at higher temperatures, the probability of molecules being in an excited state, which is given by the Boltzmann distribution, rises and scattering can also occur from the excited to the ground state.

An example of a sophisticated Raman microscope – which is in this case also combined with an atomic force microscope (AFM) in order to allow surface property analysis, is given in Figure 8.8.

Fig. 8.8: Combined Raman microscope and atomic force microscope.

Raman scattering is, compared to Rayleigh scattering, a very weak process with only one in $10^{-6}...10^{-8}$ photons showing Raman scattering. Thus, the intensities of the Stokes and anti-Stokes scattering are small compared to the Rayleigh scattering.

An often observed phenomenon is the occurrence of a fluorescence background in the Raman spectra. This fluorescence occurs when the excitation energy is high enough to promote the molecule to an excited singlet state (S_1, S_2, ..., S_n) as shown in the Jablonski scheme in Figure 8.9. This excited molecule immediately falls down to the S_1 state via internal conversion (IC). From this state, the luminescence processes fluorescence or phosphorescence may occur. In case a triplet state, obtained via intersystem crossing (ISC), is included a subsequent phosphorescence occurs. In contrast to fluorescence, phosphorescence involves a change in the spin multiplicity. Therefore, this process is relatively slow due to the forbidden S → T transition as stated by the Frank Condon principle. Thus, phosphorescence processes exhibit longer life times compared to fluorescence processes. The relative importance of fluorescence compared to phosphorescence is determined by the rates of the IC and the ISC processes [120].

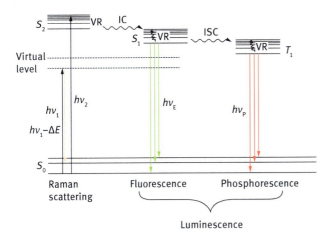

Fig. 8.9: The Jablonski scheme describes the electronic transitions of internal conversion (IC) and intersystem crossing (ISC), involving excited singlet (S) and triplet (T) states, which lead to the luminescence processes fluorescence and phosphorescence in contrast to the Raman scattering [118].

The potential fields of application of Raman spectroscopy in polymer characterization are diverse. Besides material identification, including copolymer composition and structure analysis, a conformational analysis can be carried out.

Information about the polymerisation kinetics and mechanisms can also be obtained with little sample preparation effort and also an in-situ characterization is possible. This polymerisation analysis is mostly based on the observation of the decrease in the C = C bond vibration of the reactant.

Furthermore, the degree of crystallinity of polymers can be determined via certain vibrations characteristic for the amorphous or the crystalline phase. An example is the crystallinity determination in polypropylene [121].

Different additive types are added to the polymer formulation in order to enhance processing properties, material properties of the prospective component or polymer stability. The analysis of those additives within the polymer is of great importance in order to identify the quality of the additives, their dispersion across the polymer and so on. Raman spectroscopy allows the simultaneous identification of various additives in parallel. By using Raman microscopy, even the spatial dependency of the additive distribution can be obtained [122].

Additionally, Raman spectroscopy can be used to monitor polymer degradation in a nondestructive and quick manner. Since most polymer degradation reactions can be traced back to thermo- or photo-oxidative processes, this Raman spectroscopic degradation analysis is often the observation of the occurrence of peaks of oxidised species [123].

In order to analyse full-size PV modules, a Raman probe can be attached to the Raman spectroscope. In the example given in Figure 8.10, the Raman probe is attached to a linear motor which allows an automatic movement of the probe and so also reproducible measurements.

Fig. 8.10: A Raman probe which is attached to a linear motor.

8.2.6 Nanoindentation

A nondestructive method for the determination of the degree of cross-linking in EVA solar module encapsulation is the nanoindentation test, as shown in Figure 8.11. The technique relies on measuring the shear storage modulus of the encapsulation material by using a standard DMA instrument. Some groups found a good differentiability of samples with a range of different gel contents, and the power law relationship between shear modulus and gel fraction was found to agree with rubber elasticity and swelling theory [124].

Fig. 8.11: Prototype of a indentation tester. Source: Fraunhofer CSE.

The problem of this method is the necessity of having a nonaged sample as reference on which the measurement of the aged sample can be referred to. Another disadvantage of the technique is that you can only measure the rear side EVA and only at modules which have a polymeric back-sheet as you measure through the back-sheet [125]. Additionally, a potential aging going along with altered elastic properties of the back-sheet could also lead to incorrect gel content values.

8.2.7 Fourier-Transform FT-IR/UV/vis measurement

The transmission and reflection properties of PV module components or small PV laminates can be quantified by means of Fourir transform FT-IR/UV/vis spectroscopy.

The absorption of light is described by Lambert–Beer's law:

$$A_\lambda = \lg(I_0/I) = \varepsilon_\lambda cd, \quad (8.7)$$

where A_λ is the absorbance, I_0 is the intensity of the incident light, I is the intensity of the transmitted light, ε_λ is the extinction coefficient, c is the concentration of the absorbing species, and d is the path length.

Since the absorbance of a sample is related to the transmittance T_λ via an inverse logarithmic dependence

$$A_\lambda = \lg(1/T_\lambda), \qquad (8.8)$$

the amount of light transmitted through a certain sample can therefore be calculated according to

$$T_\lambda = I_0/I = 1 - R_\lambda - A_\lambda, \qquad (8.9)$$

where R_λ is the reflectance and of the sample.

For FT-IR/UV/vis spectroscopic measurements, light of a broad spectral range is employed for the excitation of molecular vibrations in the IR, UV and visible range. With different light sources such as xenon lamps, tungsten lamps, metal halide lamps, and SiC globars, a wavelength range between 200 and 17000 nm can be achieved. For the detection of the transmitted or reflected light, different detectors, such as photo multipliers, silicon detectors, InGaAs detectors, and mercury cadmium telluride (MCT) detectors, can be used.

The measurement of a larger measurement spot is used due to an often inhomogeneous nature of samples. Such larger measurement spots can be accomplished by integrating spheres (Ulbricht spheres) which are shown in Figure 8.12. Ulbricht spheres consist of a hollow spherical cavity with a diffuse reflective coating, such as Barium sulphate for the UV/vis or gold for the IR range, and small holes for entrance and exit ports. Light rays which incident into those integrating spheres are distributed equally to all other points. Therefore, an integrating sphere is a diffuser which preserves energy but destroys any spatial information.

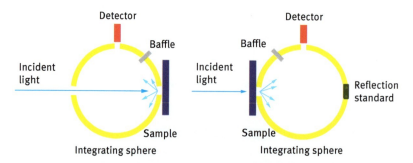

Fig. 8.12: Integrating spheres are either utilised in reflection (left) or transmission mode (right). The incident light is in both cases scattered by the diffuse reflective coating on the inside of the sphere. The detection of the scattered light specific is achieved via detectors for the UV, the visible and the IR range.

The specific detectors for the UV, the visible and the IR range are situated on top of the sphere. The measurement can be performed in a transmission mode (right) or reflection mode (left), depending on the mounting of the sample behind or in front the sphere. Integrating spheres can, among other things, be utilized for the measurement of the radiation characteristics of a lamp or to measure the diffuse reflectance of surfaces, including an average over all angles of illumination and observation.

8.2.8 Scanning acoustic microscopy SAM

Scanning acoustic microscopy is a relatively new technique which uses focused sound for the investigation of a sample. This way, voids, cracks, and delaminations within microelectronic packages can be detected in PV modules. The technique is based on acoustic signals from the sample. Therefore, a microscope produces, transmits and detects short pulses with a high penetration rate. Changes in the partial transmission and reflection of ultrasound can be detected in case of changes in the acoustic impedance, which occur if material boundaries or property changes are present. Frequencies up to 2000 MHz are used for the detection of potential problems within the laminate.

9 Loads for PV Modules

Like for all technical products operating outdoors, the materials used in PV modules are exposed to loads influencing their performance and causing degradation effects. Since for PV modules service life times of 20, 25, or even more years are expected the investigation of degradation effects, and relevant loads and stress factors is of special importance.

The loads can be categorized into external loads by weather and internal loads which can be related to the operation of the PV module itself. This chapter lists the most important load factors and describes their most relevant properties and influences.

9.1 External loads

Besides the materials, the design and the processing, PV module degradation is mainly influenced by the external loads. As visualised in Figure 9.1, these loads are

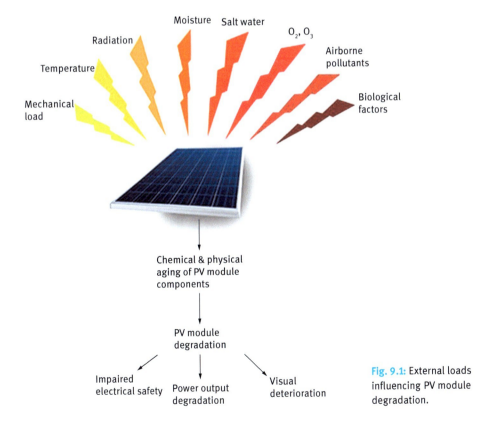

Fig. 9.1: External loads influencing PV module degradation.

mechanical loads, temperature, radiation, moisture, salt water, as well as chemical and biological loads. The influence of these factors may cause chemical and physical degradation processes on the inside of a PV module as well as on its surface.

9.1.1 UV-radiation

The electromagnetic radiation coming from the sun is called global irradiation or solar irradiation. Its spectrum and intensity depends on the geographical location, the altitude, time, date, and weather at the specific place of measurement. The terminus air mass (AM) is used to describe spectral differences due to location and time. The spectrum of the irradiation outside the atmosphere is called AM 0, and the spectrum of irradiation after crossing the atmosphere and reaching the earth surface on the shortest way is called AM 1. In the PV industry the spectrum at AM 1.5 is commonly used as reference spectrum for measurement of PV modules performance and solar simulators since it describes the solar spectrum reaching the ground in central Europe (exactly 48.2° north) around noon on a clear day. With a peak intensity of 1000 W/m² it is described in the international standard IEC 904-3 (1989) part III.

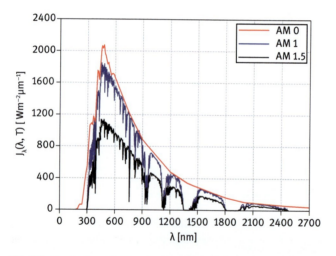

Fig. 9.2: AM 0, AM 1, AM 1.5: solar spectra outside the atmosphere (AM 0), at vertical incidence at the equator (AM 1) and reverence spectrum for solar application with zenith angle of 48.2° (AM 1.5).

Radiation can cause different effects when it is absorbed by materials. On the one hand there are positive effects like the generation of electric power in PV cells, but on the other hand there are also numerous negative effects causing degradation. The absorbed radiation can raise the temperature of the material leading to temperature dependent effects but it can also directly cause damage. Each molecular bond has a characteristic energy which is necessary to break it. Thus, photons with an energy which is at least as high as this dissociation energy E_D can destroy the chemical bonds and cause changes of the material and its properties. In PV modules especially the polymeric components are sensitive towards solar irradiation. The energy which is necessary to destroy the polymeric chains usually requires photons of the ultraviolet (UV) part of the solar spectrum. Therefore this part of the solar spectrum is of specific relevance when dealing with reliability issues of PV modules. As the longer wavelength parts of the light influcence the temperature of the materials and the measurement of UV-irradiation is neither cheap nor (usually) precise, usually the intensity and dose of the full solar spectrum is measured.

For the calculation of the UV dose, very often an estimation of 5.5 % UV compared to the full intensity, which is based on measured doses of the global irradiation, is used. This approximation can be utilised for the estimation of the UV loads.

9.1.2 Temperature

The temperature has a significant impact on most chemical processes and is thus very important with regard to degradation processes. The influence of temperature on the speed of reactions, i.e. its role as accelerating factor, can be described by the Arrhenius equation (equation (9.1)):

$$k = A \cdot e^{\frac{-E_A}{R \cdot T}}, \tag{9.1}$$

with

A   material factor,
E_A activation energy [J· mol^{-1}],
R   8, 314 J · K^{-1} · mol^{-1} – universal gas constant,
T   absolute temperature [K],
k   reaction rate.

The temperature can be determined via reliable and fast measurements with cheap sensors like thermo elements or PT-100 resistivity sensors. As temperature data is available for many locations, realistic real-life stresses can be used as a basis for aging tests. Here, the heating effect influence of solar radiation on materials in PV modules, as described in the "UV" section, have to be taken into account. Thus, significant differences between ambient temperature (macroclimate) and sample temperature (microclimate) have to be expected. Therefore, the microclimate around a sample should be regarded as the relevant one for degradation effects. The differences between ambient and module temperature can be as high as 30 K as shown in Figure 9.3. In addition, the problem of a reliable temperature determination of surfaces under irradiation has to be dealt with. By using inadequate measurement set-ups or sensors, misleading temperatures can be determined. Thus, temperature sensors should be selected and placed onto samples with great care.

Next to the acceleration of chemical reactions, also mass transport reactions are accelerated at high temperatures. Thus, at higher temperatures more water vapor or oxygen from the atmosphere can ingress into the module and cause hydrolytic or oxidative reactions. At the same time, degradation by-products can also outgas out of the module, which may reduce the corrosive stress within the module laminate [126].

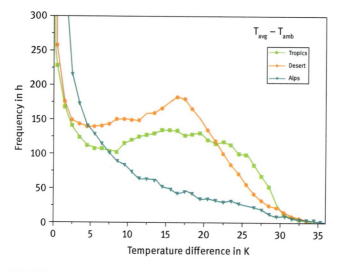

Fig. 9.3: The module temperature can vary dramatically from ambient temperature, depending on the climatic conditions and the absorption properties of the PV module [127].

9.1.3 Humidity

As a certain amount of humidity is available in all climates at all times and therefore has to be taken into account when it comes to exposure of PV modules, especially since its influence on degradation effects is very strongly dependent on the materials used.

While for coatings often the condensation of humidity on the outer layers due to dewing is of major relevance, polymeric degradation effects like hydrolytic degradation processes occur on the inside of the module. Thus, the humidity level has to be determined with respect to the relevant effect of humidity. While the measurement of the relative ambient humidity is easy and only requires simple sensors, the relevant value or the degradation of samples is the humidity level directly at or in the material (microclimate). So the use of ambient data for the samples is usually not possible. The necessary direct measurement on the surfaces or in the material itself is often not possible or also influenced by artefacts, e.g. due to the humidity capacities of the sensors themselves and the low absolute humidity levels in encapsulation materials. Usually the best way is the calculation of the microclimatic humidity based on ambient humidity and material temperatures. For theses calculation the temperature is of major relevance since the humidity capacity of most materials strongly depends on the temperature, especially the humidity capacity of air. Therefore for objects with a higher surface temperature as the ambience like PV modules in operation the relative humidity is much lower than in the measured ambience. On the other hand for objects which are colder than the ambience, like PV modules, during clear nights the humidity is higher, which can cause condensation.

The definition of the critical humidity level above which the humidity is relevant for degradation processes or above which the sample can be seen as humid or wet is often also not easy or not possible. Therefore values like the "time of wetness (TOW)" are used to quantify times above a certain level of humidity and compare locations. Times with relative humidity > 80 % are often counted as TOW, but this level strongly depends on the used materials.

Since exterior PV module components are chosen to be humidity stable, atmospheric humidity mainly impairs the interior PV module components. The water vapor diffusion proceeds from the atmosphere through the back-sheet foil and the encapsulation layer up to the intercell connectors, the cell metallization, and the solar cell. All those metallic components can subsequently undergo corrosion reactions. But hydrolysis of the encapsulation material or the back-sheet core layer can also lead to an impaired PV module performance or electrical safety. Thus, the permeation processes which proceed in a PV laminate are crucial for understanding the humidity-induced PV module degradation.

Permeation is a mass transport process of molecules through a membrane from an exterior to an interior environment, which is accomplished by diffusive processes.

Chemical reactions and diffusive processes are accelerated by increasing temperatures according to the Arrhenius equation, which has been described at the beginning of this chapter. The influence of this temperature dependence of the water up take in the encapsulation material in a PV module is displayed in Figure 9.4.

The water ingress into the front-side encapsulation layer, up to the centre of the solar cell, over a period of 20 years, was simulated for two different back-sheet materials (PET and PA) and four climates (tropic, alpine, arid and moderate). The results show significant differences in the time-dependent water ingress, mainly due to the differences in relative humidity and ambient temperature at the different exposition sites, but also by the permeability of the back-sheet materials. The length of time of the initial accumulation phase before equilibrium state was found to be highly dependent on the ambient temperature. For cold climates (alpine), the accumulation time takes about ten years, while it only takes two to three years in hot, arid climates. In contrast, the actual amount of water present inside the PV module is mainly determined by the humidity of the surrounding atmosphere. Therefore, water concentrations in humid climates (tropic, moderate) are up to 100 % higher than in hot and dry climates (arid). The reason for the counterintuitive relationship between the water content in moderate and tropic climates is the outgassing of previously incorporated water vapor

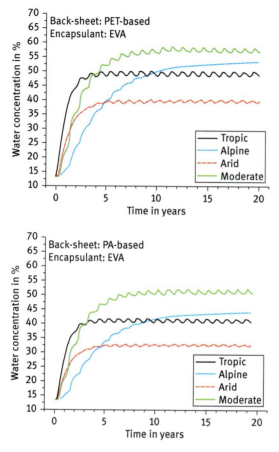

Fig. 9.4: The simulated water ingress into a PV module for two different back-sheet materials (PET and PA) and four different climates (tropic, alpine, arid and moderate) displays significant differences in the time-dependent water ingress, which is influenced by the relative humidity, the average temperature and the permeability of the back-sheet [128].

which proceeds more efficiently in climates with high daytime temperatures. The water uptake is also influenced by the permeation properties of the back-sheet, caused by different temperature dependencies of different back-sheet materials, as shown in Figure 9.5.

These different temperature dependencies of the WVTR result in differences in the acceleration of both water vapor uptake into the module and the outgassing of water vapor. Therefore, e.g. for a module with a PA back-sheet less water vapor can ingress into a module during cold (night) periods, and more water can outgas during periods of high module temperatures, which occur mostly during daytime, when the module is hotter than the environment due to solar irradiation, compared to a module with a TPT back-sheet.

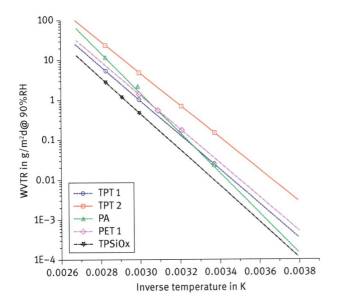

Fig. 9.5: The measured water vapor transmission rates (WVTR) of several back-sheet materials display significant differences with regard to their temperature dependence [128].

9.1.4 Mechanical loads

Mechanical loads on PV modules are usually caused by
- wind,
- rain,
- snow,
- hail,
- ice.

The wind load impairing on a PV module stems from the pressure distribution around it.

Besides an area of overpressure in front of the module and of negative pressure behind the module, turbulences are induced and as a result, a displacement of the module perpendicular to the plane of mounting is caused. This displacement results in a mechanical stress which is most harmful for the cell interconnectors and the cells and may therefore lead to a fatigue of the material and to the formation of cell cracks. Of the listed precipitation impacts rain, hail, snow and ice, hail is the only one which has an instantaneous impact on the module. In dependence of the respective hail grain size, hail may lead to glass breakage and in turn to an impaired electrical safety and stability of the module. While rain droplets can induce vibrations inside the PV module, snow and ice, accumulating on the surface of a PV module, can cause high static mechanical loads of up to several thousand Pascal which may lead to delamination, glass and cell breakage, or the detachment of the frame.

9.1.5 Corrosivity

Besides polymer degradation, the degradation of the inorganic components in a PV module is one of the most important aspects of PV module degradation [129]. Significant decreases in PV module performance are caused by the corrosion of the cell, i.e. the SiN_x AR coating or the corrosion of the metallization, i.e. solder bonds and the Ag-fingers [111, 130–132].

As stated above, metal corrosion is an electrochemical process, which is the reason for the necessity of an aqueous electron conduction-enabling environment to be present. Corrosion is a surface process in which atoms at the surface of the metal enter the solution as ions and at the same time, electrons migrate through the metal to a site where they are consumed by species in contact with the metal [133]. The metal ions may also combine with other species in solution and compounds such as oxides or hydroxides can precipitate.

A corrosion process takes place spontaneously when the change of the free enthalpy ΔG during this process is < 0. For an electrochemical reaction, ΔG is proportional to z, which is the number of electrons transferred in the reaction, as well as to the difference ΔE between the standard potentials ΔE_{red} and ΔE_{ox}:

$$\Delta G^\circ = -zF\Delta E^\circ. \tag{9.2}$$

The external conditions can be integrated in this relationship via the Nernst equation

$$E_{red} = E^\ominus_{red} - \frac{RT}{zF} \ln \frac{a_{red}}{a_{ox}}, \tag{9.3}$$

where E_{red} is the half-cell reduction potential at the temperature of interest, E_{red} is the standard half-cell reduction potential, R is the universal gas constant, T is the temperature, a the activity for the relevant species, and F the Faraday constant. In a PV module the front side cell metallization, i.e. the grid which mainly consists of silver, as well as the Sn/Pb capped copper intercell connectors and the rear side metallization, which is typically made from aluminum, are the corrosion susceptible components. The necessity of an electrolytic environment is met when water vapor diffusion into the module, i.e. in to the encapsulation creating an ion conduction environment, is possible. The corrosion kinetics are either determined by the through-passage of the ion or by the diffusion of the electrons. Besides the silver grid on the silicon cell, the intercell connectors, can undergo corrosion reactions.

9.1.6 Other chemical loads

The presence of oxygen is crucial for all oxidation reactions, which constitute the most important polymer degradation mechanism. Similar to humidity, oxygen has to permeate through the back-sheet material in order to initiate oxidation reactions in the encapsulation material. The oxygen transmission properties of different back-sheet materials differ extremely for the different materials. The measured oxygen transmission rates (OTR) of several different materials can be found in Figure 9.6.

The presence of oxygen limits all oxidative reactions, and therefore a high oxygen permeation rate can result in an acceleration of the polymer oxidation reactions, which can result in chain scission and cross-linking reactions.

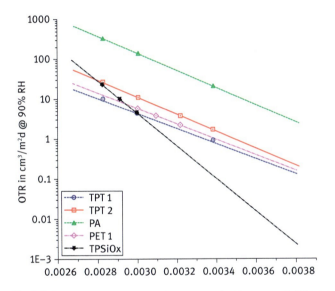

Fig. 9.6: Measured oxygen transmission rates (OTR) of several different materials [132].

In addition to oxygen, other gases which are present in the atmosphere or which can be formed under certain conditions can ingress into PV modules and cause degradation reactions or also lead to a degradation of the external PV module components such as the frame. Gases like NO_x, SO_x, and O_3 can be washed out by rain events and precipitate on the surface. In the presence of water, nitric acid and nitrous are formed from NO_x, and sulphur acid as well as SO_3 come from SO_x. Those acids can cause hydrolytic bond cleavage of certain polymers like polyamide. Also under the impact of dew and fog, minimal pH values of 1.5 were measured in Germany. The influence of corrosive gases on the PV module degradation has been to date hardly investigated. High levels of chloride deposition can be found in coastal regions, but the chloride content in the air decreases rapidly with increasing distance to the shore line. Chloride ions can move through the layer of corrosion products and arrive at metal surfaces where they accelerate the corrosion process [134]. For the anions in this process, a catalyst role can be considered [135]. Sulphur dioxide (SO_2) is mainly released as side product from the combustion process of oil or coal due to a certain sulphur content. Therefore, main sources of SO_2 pollution are energy plants and car exhaust fumes. Thus, SO_2 levels are increased in industrial environments. In the past two decades, the usage of low-sulphur fuels resulted in a dramatic decrease of the SO_2 in the atmosphere. The solution of SO_2 in water may result in the formation of sulphur acid, which accelerates metal corrosion. The SO_2 content is determined via the SO_2 deposition rate or per concentration measurements. Figure 9.7 shows the distribution of atmospheric corrosivity on the global map was created using a geographical information system [136]. Thus, the chloride deposition is only important in areas less than 500 m away from the sea [137]. Since sea salt, i.e. NaCl, is a strong electrolyte when diluted in water, reactions requiring conduction as well as ion exchange on the surface of metals are accelerated under the presence of Cl^-. The amount of chloride in the atmosphere is determined by the chloride deposition rate, which can be measured either by the "wet-candle" method or by conductivity

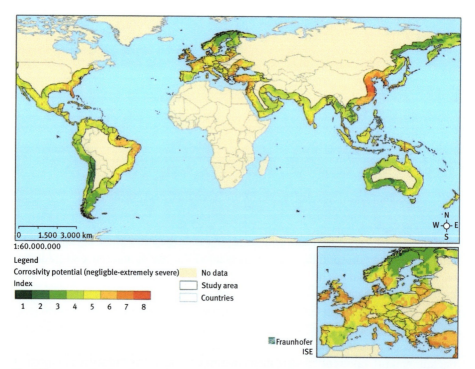

Fig. 9.7: The distribution of atmospheric corrosivity on the global map was created using ArcGIS considering the chemical factors salinity, SO_2 pollution, and relative humidity of the atmosphere [136].

Atmospheric corrosion in coastal regions can be classified into eight groups according to the key corrosion factors: relative humidity, airborne salinity, and sulphur dioxide pollution. A corrosivity map with a color code ranging from green (very mild atmospheric corrosion) to dark red (extremely high corrosion rate), shows that the atmospheric corrosivity is very heterogeneously distributed around the world. In some regions, such as China and Brazil, the Arabian peninsula, or the east coast of the US, high atmospheric corrosion can be found. This high corrosion levels can be attributed either to very high airborne salinity or to the interaction of the factors humidity, salinity, and air pollution. Especially eastern China with its high atmospheric corrosivity is a good example for this synergetic effect [136].

The degradation of metals by airborne pollutants was attributed to the deteriorative of sulphur dioxide and in special environments also to chlorides. Therefore, those pollutants received thorough attention in atmospheric corrosion research and have been most extensively studied. Environmental protection laws enacted about two decades ago in industrialized countries led to a notable decline in the sulphur dioxide levels. At the same time, concentration levels of other air pollutants, such as the nitrogen oxides, and secondary pollutants, such as ozone, however, remained rather constant or even increased slightly. This has increased the relative significance of these air pollutants in the atmospheric degradation of materials [138].

10 Accelerated aging tests

All environmental simulation tests which allow an accelerated analysis of material degradation processes through intensified environmental stress levels compared to real life can be summarized under the notion of accelerated aging tests.

This type of test is especially useful for products with long service lifetimes, such as photovoltaic modules. As guarantees for PV modules cover up to 30 years of service life, accelerated aging tests are of high economic importance for PV modules manufacturers. As PV manufactures aim for utilizing materials which are both economically and technically suitable, accelerated aging tests on full PV modules as well as on separate materials and material combinations are necessary. It is in the interest of the customer as well the investor and the insurance that the product achieves the expected service life without major performance losses. The aim of accelerated aging tests is to achieve reliable results within a very short period of time and with minimum costs in order to keep pace with the rapid innovation cycles. In order to ensure the transferability of the testing result on real life performance, a specific adaptation of the test conditions to the specific materials and stress factors is indispensable. Otherwise, climatic conditions may be induced which either exceed or undermine those under standard operation, and this may subsequently lead to altered physical or chemical degradation processes. Thus, the processes which occur under standard operation should be accelerated but at the same time simulated as close to reality as possible.

The acceleration factor a is thus

$$a = \frac{t_{test}}{t_0}, \qquad (10.1)$$

with the testing time during accelerated aging (t_{test}) and exposition time under standard operation (t_0), which both result in an identical material degradation. The dependence of the acceleration factor on the testing conditions and the degradation factors are described by time transformation functions which can be used for the estimation of the service life of a product. In general there are three ways to achieve the required acceleration:
1) constantly high stress level instead of a variation of the intensity of the degradation factor;
2) increase of the intensity of the stress factor;
3) increase of the temperature for the acceleration of the degradation process.

In all cases, the potential of inducing unrealistic processes and subsequently unrealistic material degradation processes exists. By 1) relaxation phases which may occur in reality, e.g. during the night, or times needed for transportation processes can be neglected. Option 2) has the risk of altering the dose response relationship significantly, e.g. if the limiting reaction process has not enough time or reaction partners to react. Possibility 3) bears the danger the different processes have different temperature-dependencies, and thus an alteration of the reaction equilibrium may occur. Additionally, temperatures may be too high and this way, non-realistic conditions may be achieved. Therefore, the specific conditions should be evaluated thoroughly before choosing the most suitable conditions.

10.1 Light sources

As described above, the degradation process of PV modules depends highly on the irradiation level during the aging process.

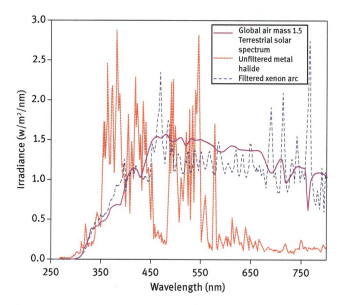

Fig. 10.1: Spectral distribution of the irradiance of the sun (AM 1,5 Spectrum, magenta), of a metal halide lamp (red) and of a xenon lamp with a filter (blue) [139].

Ideally, an accelerated aging test would use the full solar spectrum, e.g. at AM 1,5-spectrum as shown in Figure 10.1, in order to eliminate issues related to altered spectral distributions during the aging process. But due to the occurrence of an overheating of the sample, especially in case of high irradiance levels, as well as the predominant impact of short wavelength irradiation on the degradation processes, light sources are usually chosen according to their UV levels. As the chemical bonds in the polymeric materials in PV modules are especially susceptible to certain ranges of UV irradiation, it is very important to understand the spectral sensitivity of the materials. The energy of the photon is of great importance regarding the radiation's impact on the polymer. If sufficient energy in form of radiation or heat is supplied, an initial chain scission step can occur, followed by the formation of two alkyl radicals. Without the necessity of any further energy input, those radicals may react with molecular oxygen to form a peroxide radical. In the case of additional energy input, the peroxy radical can abstract a hydrogen atom from an alkyl chain, resulting in a hydroperoxide and an alkyl radical which both can react with oxygen to form a peroxide and a peroxy radical, respectively. This way, hydroperoxydes are generated which may decompose to an alkoxy and a hydroxyl radical. Those radicals may break further C-H bonds under the formation of alcohol or water and alkyl radicals. The latter ones in turn react with oxygen to peroxy radicals.

The efficiency with which an incoming photon may induce damage in a material varies between different materials, depending on colour, surface conditions, and other factors, but is exponentially related to the wavelength of the incident photon.

The energies and corresponding wavelengths which are required to induce bond breakages of the most important organic groups are given in Figure 10.2.

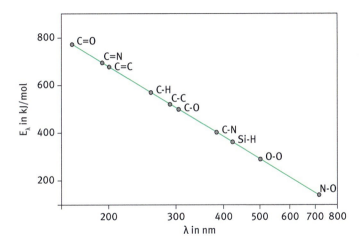

Fig. 10.2: The energies and corresponding wavelengths which are required to induce bond breakages of the most important organic groups, according to [140].

Different types of light sources can be used for accelerated aging tests. These light sources fulfil the described requirement to different extends. In case the spectral sensitivity of the sample is unknown, it is suggested to use a light source with a spectral distribution close to the solar spectrum. Xenon light sources with filters are especially suited for this application as shown in Figure 10.3.

For the analysis of polymers it is usually important to achieve very similar sample temperatures between different samples while at the same time, the UV range of the spectrum is aimed to be enhanced. For this objective, light sources with a relatively high UV portion such as metal halide lamps or fluorescence tubes are most suitable

(Figure 10.3). For choosing the most suitable light source for a UV test is highly important to now the spectral distribution as the applicable norms such as ISO 4892, IEC 61215, 10.10 allow for a broad range of possibilities as shown in Figure 10.3.

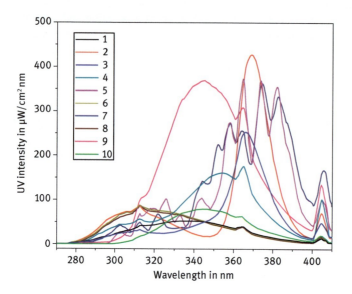

Fig. 10.3: Comparison of the intensity of the spectral distribution of 10 different UV light sources used in a round robin between different test laboratories [141]. The light sources are UVA fluorescence tubes (1, 9, 10), UVA+UVB fluorescence tubes (2, 3, 4, 6, 8) and metal halide lamps (5, 7).

For the analysis of large samples, the spectral and spatial homogeneity of the light source is of great importance in order to allow for a comparability of the results. The temporal and spatial homogeneity of the spectrum of solar simulators is for instance classified in norm IEC 60904-9. These issues are usually easier to implement with fluorescence tubes compared to point light sources.

10.2 Climatic chambers

Climatic chambers are all devices which are able to produce and maintain specific environments for accelerated aging tests of test samples. The simplest form of climate chamber is a climatic chamber which maintains a certain temperature level, also referred to as an "oven". Two examples of these "ovens" which are suitable for very different temperature ranges are shown in Figure 11.1 in Chapter 11. More sophisticated devices allow for an additional humidity regulation inside the test cell. An example for such a climatic chamber, which is used for the standard PV module damp-heat test is displayed in Figure 10.4. In case of the additional possibility of adding radiation load

Fig. 10.4: Climatic chamber which allows the regulation of temperature and humidity inside the test chamber.

to the accelerated aging test, the chamber can also be referred to as weathering testing instrument. An example is given in Figure 10.5.

Within this climatic chamber type, the exact regulation of the climatic conditions is rather difficult, as the light sources result in an intense heating of the chamber, which in turn needs to be extracted. Generally, the homogeneity of the chamber con-

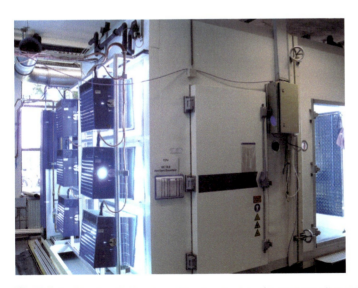

Fig. 10.5: Double climatic chamber with solar simulator (class B according to IEC 60904-9).

ditions throughout the whole chamber is a critical point, especially in case of rapid temperature cycling and at high temperatures and high humidity and high irradiation.

As described in Chapter 9, homogeneity with regard to time and space is often a critical point during accelerated aging tests. Additionally, it is important to differentiate between the ambient atmosphere, which is the given environment of the test chamber, and the microclimate, which is the resulting environment directly at the sample. As the parameters of the environment are amplified compared to real-life conditions, the resulting microclimate is also often significantly altered, which may lead to a significant acceleration of some of the degradation processes. While this is a generally desired phenomenon in order to achieve short testing times, it is of great importance to know the exact amount of amplification. Thus, precise measurements of the different parameters of the microclimate are absolutely crucial for a successful and reliable accelerated aging test. The determination of the sample temperature is thereby most difficult during an accelerated aging test, which includes irradiation of the sample. To address this problem, the so-called black standard temperature (BST) or white standard temperature (WST) are available. The optical properties and the related surface temperatures usually lie in between the two extremes of the BST and the WST. The direct measurement of the sample temperature can also be difficult, as the measurement itself can impact on the conditions at the sample [142]. The microclimate

Fig. 10.6: Salt mist chamber at the Fraunhofer ISE.

should be considered especially in the case of several samples in one test with the aim of comparing the results,

Corrosive environments have a major impact on the degradation PV modules, as described in detail in Section 9.1. Therefore, accelerated aging tests which simulate the degradation characteristics in marine or other high-corrosivity environments are of great importance, even though they are relatively new compared to established tests like UV or damp heat aging. As salt mist is one of the most important factors for this type of degradation, salt mist chambers as shown in Figure 10.6 can be used for accelerated aging. Other corrosivity tests can be performed specific corrosive gas chambers, such as e.g. ammonium test chambers.

10.3 Procedures

The amplification of the different loads during an accelerated aging test results in various risks and difficulties, which were described earlier in this chapter. These risks have to be taken account in the design of an accelerated aging test procedure. Therefore, a comprehensive experimental design is required. This experimental design has to consider all available information with regard to the sample, such as specific temperatures which should not be exceeded or temperatures at which physical transitions (T_G, T_M) occur, or known interactions with other materials which will be present in the final application.

On this basis, with the help of a statistical experimental design, a comprehensive experimental plan can ideally be developed which aims at evaluating the dose response relationship between the different parameters and degradation indicators with as few as possible separate aging tests. Since every single parameter has to be varied at constant conditions of the other parameters, a multitude of different climatic conditions, and in turn a large number of separate aging tests arise. This may easily lead to problems with regard to time and costs. Therefore, this ideal statistical experimental design often has to be simplified and screening tests for a sensitivity analysis are a feasible way to do so. An example of a set of parameters for an experimental test sequence for polymeric materials is given in Figure 10.7.

These tests aim at identifying the sensitivity of the samples with regard to specific loads or combinations of loads, as well as at determining thresholds which should not be exceeded. This procedure usually starts off with climatic conditions of well-established accelerated aging tests for the type of sample investigates – such as the damp-heat test for PV modules at 85 °C and 85 % r.h. – followed by a variation of the single parameters. In the case of a planned utilisation of the aging test results for modelling it is of great importance to obtain enough data points in order to achieve a reliable basis for the calculation of the kinetics and the dose-response relationship.

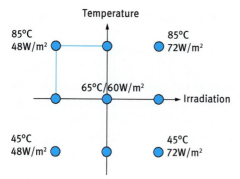

Fig. 10.7: Exemplary experimental design with a variation of the parameters temperature and irradiation.

The selection and the adjustment of specific parameters of the accelerated aging tests with regard to the expected real life conditions during the objective service life and application is called test tailoring.

11 Reliability testing of materials

In the following, some of the most important equipment items for the reliability testing of nonlaminated materials for photovoltaic applications are presented. It must be mentioned that for reliability testing, it is important to know the real occurring loads for the tested materials as well as the relevant operational conditions. This is also the major difference to accelerated ageing testing which is in the first approach independent of the operational conditions even when the equipment is similar. Therefore the specific tests in reliability testing can be very similar as in the accelerated aging tests, but the definition and the interpretation are more difficult and require more information, and thus the results are also much more meaningful for the application. In the following, a general procedure of qualification testing of materials is given.

11.1 Equipment

In Chapter 9, the general importance of temperature for the durability of materials was described. A very basic and simple heat stability testing can be performed in ovens. Figure 11.1 shows the ovens for material aging which are suitable for accelerated aging tests up to 200 °C (left) and up to 800 °C.

Fig. 11.1: Ovens for material aging which allow accelerated aging tests up to 200 °C (left) and 800 °C (right).

In addition, climatic chambers which allow heat, humidity, and UV testing – as described in Chapter 10 – are also used for reliability testing of materials for PV applications. Since for reliability testing, combined loads often have to be simulated, e.g. combined exposure to UV loads, temperature loads, and humidity loads, more flexible and sophisticated and multifunctional equipment, like climatic cabinets, with the possibility to also apply UV loads, are often necessary.

As described in Chapter 9, the water vapor and oxygen permeation through polymeric materials in PV modules is of great importance for the durability of the whole module. In order to test the specific permeation properties of back-sheet and encapsulation materials special test chambers were developed which allow a time- and temperature-dependant measurement of the permeation rate. In Figure 11.2 such a chamber, which utilizes a mass spectrometer as a detector, is shown. This is not a reliability test by itself but the setup which delivers relevant data e.g. of the backsheet and the encapsulant, for the definition of test parameters, especially for materials in the module which are separated from the ambience by the barrier material like the encapsulant or the backsheet.

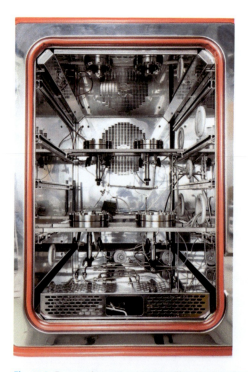

Fig. 11.2: Permeation measurement climatic chamber which utilizes a mass spectrometer as detector.

11.2 Procedures

Figure 11.3 shows the general procedure of reliability testing of new materials for PV applications which is often used for the qualification of materials during the development process. This procedure involves an initial stage in which the new material is characterized, e.g. my means of transmission measurements, tensile testing or Raman spectroscopy, dependant on the material to be investigated. These new materials are then subjected to different accelerated aging tests, such as damp-heat testing, UV testing, or combined UV and damp-heat aging, followed by a characterisation routine identical to the one before the aging procedure, in order to compare the material characteristics before and after aging. In case the materials show an acceptable aging behavior in this initial stage, small PV modules are manufactured which include not only the new materials but also all other PV components in a standard type manner. This allows testing for the specific materials interactions which are to be expected in a full size PV module. Especially the material interactions between the solar cell metallization, the encapsulation, and the back-sheet material are of great importance,

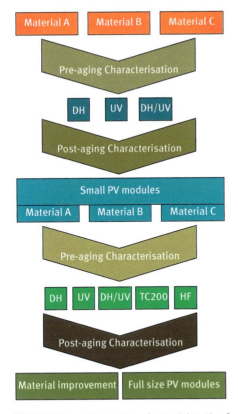

Fig. 11.3: General procedure of material testing for PV application.

as e.g. corrosive degradation by-products may dramatically change the degradation pathways and thus the speed of the degradation. Therefore, it is important to ensure that realistic microclimatic conditions occur in the weathering tests for the material of interest, and an over-simplification of the test laminate design should be prevented. The small PV modules (or test laminates) are subjected to more comprehensive accelerated aging tests such as humidity-freeze and thermal cycling, which do not have to be included in the initial phase, as these tests are more adequate when solar cells are present. These tests can also be adapted to the expected climatic and operational conditions for the use of the materials; for example, materials for roof integrated modules in southerly countries should be tested at higher temperatures than materials for the application in open rack installations in Germany. Depending on the materials, reliability tests can also be performed on single material samples, and mini modules are not necessary if the functionality is not influenced by the interaction with other components like the coatings of the glazing material. Also, the materials in small modules undergo a pre- and post-aging characterization routine which may include analytical tests as described in Section 8.2. Depending on the outcome of these tests and characterisations, the second phase will either be followed by material improvement activities or, in case of well-performing small modules, full size PV modules can be manufactured and tested.

12 Reliability testing of modules

12.1 Equipment

The general equipment which is used for reliability testing of PV modules is identical with that described in the accelerated aging chapter (Chapter 10) and reliability testing of materials (Chapter 11). As a matter of fact the equipment is required to be bigger in order to ensure that full-sized modules can be tested; this can be a challenge if e.g. irradiation, humidity, and temperature are to be controlled simultaneously.

12.2 Procedures

The major difference between accelerated aging and reliability testing lies within the general approach to the testing procedure. In contrast to the load-based approach during accelerated aging tests, which aim at simulating the various loads occurring under real life conditions separately or in combination, reliability testing aims at phenomenological specific degradation phenomena which tend to occur under operation. Therefore it is even more important for the reliability testing of modules, compared to that of materials, to ensure that during the tests realistic conditions occur, especially for the microclimatic conditions for the materials within the module. This does not mean that the tests cannot use exceeding loads in comparison to real operation to accelerate the tests, but rather that the conditions have to be chosen so that no effects occur which are unrealistic for field exposure. Some of these phenomena which are reproduced in modern state-of-the-art reliability testing are potential-induced degradation (PID), snail tracks, or discoloration. A detailed description of these effects is given elsewhere [143–146]. In Figure 12.1, the schematic description of a reliability testing procedure is shown. Mandatory tests for characterization, such as power measurements, visual inspections, insulation, and wet leakage tests (all described in standards like e g IEC 61215), electroluminescence, and – when more comprehensive characterisation is required – Raman spectroscopic or fluorescence measurements should be performed after every test cycle. Optional characterizations can also be performed after each aging sequence within the test cycle. Such test cycles enable us to determine the reliability of different module types, since they expose the specimens to relevant loads and use different analytical tools to also identify changes of properties of the module or of its materials at an early stage – often much earlier than they would be recognized by power or insulation measurements. It must be mentioned that most test cycles do not include all relevant loads, due to cost and/or time limitations. Here, a careful selection is necessary in order to include the most important loads and load combinations. The rating of loads and load collectives depend very much on the module design, the used materials, and the climatic conditions as well as the type of application (e.g. roof integration vs open rack installation). The sequence shown in Figure 12.1 for example does not include external mechanical loads which would simulate e.g. snow loads or high potential loads which can cause PID effects of PV modules. This cycle focuses on chemical and physical degradation effects due to high temperatures, humidity, and UV irradiation, represented by damp-heat and damp-heat with UV tests, and internal mechanical stress, represented by temperature cycling tests. Such tests do not automatically deliver results with regard to expectable lifetime of tested module types. Since local operational conditions and climatic loads have a strong influence on degradation effects during operation, the time of service which is corresponding to e.g. one of the cycles shown in Figure 12.1 depends on these local and operational conditions. It is also absolutely important to mention that such test sequences have to be adapted to each specific module type if real service lifetime testing is required, since the rate dominating process which the acceleration factor depends on is specific to the material and material combination. The description of the design and application of service life tests would go beyond the topic of this book, especially since PV modules consist of a combination of different materials and material classes and have to bear a variety of loads and load conditions. A more detailed description of service life testing can be found e.g. in [147] or other publications focused on service life prediction.

If yields of PV modules over time are to be predicted, it is also important to analyse the behavior of the modules under conditions which do not only cause degradation of materials but can cause additional effects reducing the performance, such as for example high dust loads leading to soiling of the modules.

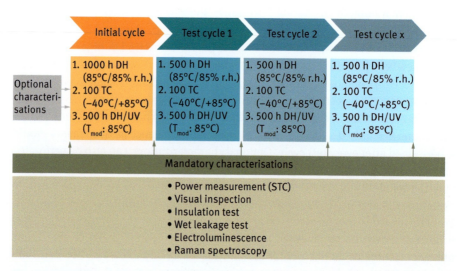

Fig. 12.1: Schematic description of a reliability testing procedure for PV modules. Specific tests are to be seen as examples.

13 PV module and component certification

Certification of PV modules and components is organized and regulated on the international level by the technical committee TC 82 solar photovoltaic energy systems of the International Electrotechnical Commission IEC. On national level so-called mirror committees exist which cover similar topics and organize the exchange between national and international level by sending representatives to IEC meetings and transfer IEC standards to national standards, so in very many countries IEC standards are translated into national standards like e.g. into European (EN) or German (DIN) standards which often use the same numbers, e.g. IEC DIN EN 61215:2011. The standards are under regular discussion to follow scientific and technical progress and industrial needs. Revisions occur and are published. This chapter describes the situation at the end of 2015. The latest versions of IEC standards are published by IEC and national bodies and can be found at https://webstore.iec.ch. At the moment it is under discussion to merge all IEC standards related to renewable energy systems under the new IEC System for Certification to Standards Relating to Equipment for Use in Renewable Energy Applications (IECRE System). This would probably cause a new scheme and numbering of standards, but this has not yet been decided, and so it is not possible to describe contents and schedule at the moment.

There is also a network of national standardization bodies and test labs which are accredited according to ISO/IEC 17025 to test according to these standards. Their test reports fulfill the requirements of the IEC System of Conformity Assessment Schemes for Electrotechnical Equipment and Components (IECEE) and are internationally accepted, so that manufacturers only have to get their products certified once.

In the following the procedure and tests for new products are described. If only minor changes in design or material selection are introduced, the full testing sequences do not have to be passed. In this case so called Retest Guidelines apply, which describe the procedure for specific situations or changes of modules. Retest guidelines can be found under www.iecee.org.

13.1 Type approval testing IEC 61215

In the 1990s an international initiative was started to develop standards for the new PV industry to assure the quality and identify products of low material or production quality. This led to a first version of the type approval standard IEC 61215, crystalline silicon terrestrial photovoltaic (PV) modules – design qualification and type approval, which was published in 1993. The latest and current version IEC 61215: 2005 is of 2005 and at the moment under revision; an updated version is expected in 2016. IEC 61215 is applicable for silicon PV modules which are used in general open-air climates, as defined in IEC 60721-2-1 on the ground, not for space applications, and also not for modules using concentrated sunlight. It is also not applicable for thin-film modules; these are covered by IEC 61646 which contains similar tests, especially with regard to weathering testing.

It is very important to mention that passing the test sequences of IEC 61215 defines a specific quality level of module design, material, and production quality, as it also includes independent inspections of the production and quality assurance measures. It does not define a specific lifetime of modules which can be expected of the certified modules. The tests of the standard focus on the identification of specific weaknesses or failures of modules, e.g. failures which can be called early failures or "infant mortality" of module types using the picture of the bathtub curve for failure rates in reliability engineering. There is neither a scientific base for relating the tests to the life span of modules nor is the purpose of the standard to answer questions related to service life, even if many people in the industry expect this or even expect a correlation of IEC testing with given long-time warranties.

IEC 61215 contains several test sequences which consist of characterization tests which measure or define a specific status of the module and also of weathering tests which expose the modules to defined loads in accelerated ageing tests. In the following the most important or crucial tests are described, a complete description of the tests and the sequences can be found in the currently valid version of the standard. Numbers are given for the different tests as they are used in IEC 61215 (10.xx), and since several of the tests are also relevant for IEC 61730, the numbers are also used there: this is the Module Safety Tests MST, and are given when applicable.

Characterization of the modules is performed in an initial state of the modules, after a preconditioning of the modules by exposing them to at least 5 kWh/m^2 of sunlight, and after the different weathering tests to identify changes due to degradation during weathering. Characterization mainly consists of the following tests:
- visual inspection 10.1, MST 01: to define the visual status;
- performance under standard test conditions (STC) 10.6_a defined in the standard, e.g. at a light intensity of 1000 W/m^2, 25 °C and an airmass AM 1.5 spectrum to define the electrical performance of the modules under defined conditions;
- electrical safety tests like wet leakage current test and dielectric resistance tests, both described in IEC 61730 below, to check safety conditions of modules.

The weathering tests and their main purpose and effects as well as limitations or weaknesses of them are described in the following:
- outdoor exposure 10.8: exposure of modules outdoors for at least 50 kWh/m^2 of sunlight, only a very low dose of sunlight, usually not enough to cause degradation or show weaknesses of materials or modules;
- UV preconditioning (UV) 10.10, MST 54: exposure of modules to 15 kWh/m^2 UV light to provoke effects due to UV ageing, e.g. of polymeric materials; UV dose only corresponds to a natural UV dose of 3 months in moderate climates like in Germany or even less in more sunny climates; the dose is usually not enough to

show the weaknesses of materials, and the test is therefore only called preconditioning; UV preconditioning is followed by TC 50 test and HF test (see below);
- temperature cycling (TC): exposure of modules to 50 (TC 50) 10.11, MST 51b or 200 (TC 200) 10.11, MST 51a cycles from −40 °C to 85 °C to provoke effects due to internal mechanical stress; causing relevant and valuable effects but no scientific correlation to outdoor loads or specific service life time of modules;
- humidity freeze (HF) testing 10.12, MST 52: exposure of modules to 10 cycles from −40 °C to 85 °C and 85 % relative humidity to provoke effects due to freezing of water or humidity which enter the modules or materials;
- damp heat (DH) testing 10.13, MST 53: exposure of modules to 85 % r.h. at 85 °C for 1000 h to provoke effects caused by high humidity and/or high temperature, like hydrolysis of polymeric materials or corrosion of glazing or coatings; causing relevant and valuable effects but no scientific correlation to outdoor loads or specific service life time of modules;
- hail impact test 10.17: exposure of modules to hail stones with defined size and speed; usually only causing effects if nontempered glass is used as glazing;
- mechanical load testing (ML) 10.16, MST 34: exposure of modules to external mechanical loads of +−2400 Pa from the front in three cycles and 5400 Pa from the front if the stability against extensive snow loads are proven; tests are performed at room temperature, which does not simulate realistic conditions for snow loads, since material properties are usually temperature dependent and cycle times do not simulate wind load conditions;
- hot-spot test 10.9, MST 22: testing of modules under illumination and partial shading of one cell to provoke hot-spots and test stability and safety under these crucial conditions.

In general one can state that the tests of IEC 61215 deliver valid and important results to identify modules of minor quality and with clear problems of materials, material combinations, or production quality, so certification should be seen as absolutely necessary if the aim is quality and reliability. But it must also be mentioned that the full type approval testing is only done with one set of eight modules. Two of them are exposed to UV, TC 50, and HF testing, two are exposed to TC 200 testing, and two are exposed to DH testing, one of them followed by ML and one followed by hail testing, one is outdoor exposed and one is kept as a control unit. Certification testing is in general (detailed pass/fail criteria are to be found in the standard, e.g. found in https://webstore.iec.ch) passed if performance after exposures is not less than 5 % below nameplate data and safety is still shown. So a total module type, often produced at different sites and over longer time periods, can be certified by testing of only eight modules. It is also important to mention that certificates loose validity if module designs or materials are changed, which is often not easy for the customer to identify. The most important point to mention is that IEC testing has to be separated form service-life testing, for which a totally different approach, including the specific designs and materials, has to be chosen, e.g. as described in Chapter 12. To sum up, testing according to the IEC standards should be seen as necessary but not sufficient to identify modules of high quality with proven reliability and durability.

Standards for type approval or qualification of materials or components like encapsulants, backsheets, glazing materials, etc., have been under discussion and preparation for several years and have reached different levels of maturity; for some materials draft versions already exist, but standards have been passed for any of them. Here it should be mentioned that for material selection and qualification, material combinations and interactions and processing are crucial, so that many tests cannot be performed or do not make sense with a single material specimen. Therefore material standards will probably only describe tests and measurement procedures, but will not be able to deliver pass/fail criteria or reach a level where they can contribute to a situation where a module of certified materials will be certified automatically.

13.2 Safety testing IEC 61730

Safety testing of PV modules as it is required by law in many countries is covered by IEC 61730. The relevance for specific countries depends on the national regulations and has to be checked for each country. IEC 61730 is divided in two parts:
- IEC 61730-1 photovoltaic (PV) module safety qualification – Part 1: requirements for construction, latest version IEC 61730-1: 2004+AMD1:2011+AMD2:2013 of 2004 with amendments of 2011 and 2013;
- IEC 61730-2 photovoltaic (PV) module safety qualification – Part 2; Requirements for testing, latest version IEC 61730-2:2004+AMD1:2011, of 2004, with amendments of 2011.

Both parts are applicable if a certification is required. Part 1 describes the general requirements which have to be met by materials and module design, e.g. distances between life parts and the frame or properties of materials. The specific values are not described here, since they are under discussion, and therefore the reader should always have a look at the latest version of the standard. Part 2 describes tests, sequences, and pass/fail criteria for safety testing of PV modules. Similar to IEC 61215 it consists of characterization tests to measure the status or changes of the samples and weathering tests. Since several tests are similar to the ones in IEC 61215, it usually makes sense to perform the tests for IEC 61215 and 61730 for one module type in parallel. Here only the most important additional requirements and tests of IEC 61730 are described. The naming and numbering of tests in IEC 61730 uses the short form module safety test (MST) for each test. Several tests are contained in IEC 61215 (numbering 10.xx) and also have a MST number, since they are part of IEC 61730. In the following the major MST tests for modules are listed and commented:

- accessibility test MST 11;
- cut susceptibility test MST 12: test of the stability of modules against mechanical impact, e.g. with tools; mainly relevant for backsheets;
- ground continuity test MST 13;
- impulse voltage test MST 14: test of electrical safety against high voltage pulses;
- dielectric resistance test 10.3, MST 16: also called high voltage test, test of the electrical insulation properties of the complete module;
- wet leakage current test 10.15, MST 17: test of the electrical insulation properties of the complete module under wet conditions, immersion of the module, spraying of junction box;
- fire test MST 23: test of the flammability of modules; mainly relevant for backsheets, performed on extras module;
- bypass diode thermal test 10.18, MST 25: test of the stability and thermal behavior of bypass diodes under load;
- reverse current overload test MST 26: test of the module behavior under reverse current load;
- module breakage test MST 32: test of mechanical integrity of module after mechanical impact, performed on extra module;
- robustness of terminations 10.14, MST 42: test of the stability and adhesion of the junction box and the cables.

IEC 61730 also describes some tests which are relevant for materials or components of PV modules to qualify these components for use in modules from the safety point of view. Here the following test is important to mention, since it directly influences the allowed operational voltage of module:
- partial discharge test MST 15: test of backsheet material to measure discharge currents.

Certification testing in the scheme of IEC with the standards IEC 61215 and IEC 61730 is an important pillar for international comparability of PV products and defining quality standards as long as all stakeholders in the value chain keep the described purpose and limitations in mind. The standards only define a specific quality level, they do not include differences between materials, material combinations, climatic conditions of the location of application etc. and all these parameters are influencing service live and reliability.

14 References

[107] E.D. Dunlop and D. Halton, The performance of crystalline silicon photovoltaic solar modules after 22 years of continuous outdoor exposure, *Progress in Photovoltaics: Research and Applications*, **14** (2006), 53–64.

[108] H. Laukamp et al., Reliability issued in PV systems – Experience and improvements, 2nd World Solar Electric Buildings Conference, Sydney, 2000.

[109] S. Kaplanis et al., Energy performance and degradation over 20 years performance of BP c-Si PV modules, *Simulation Modelling Practice and Theory* **19** (2011) 1201–1211.

[110] M. Vázquez et al., Photovoltaic module reliability model based on field degradation studies, *Prog. Photovolt: Res. Appl.* **16** (2008), 419–433.

[111] C. Dechthummarong et al., Physical deterioration of encapsulation and electrical insulation properties of PV modules after long-term exposure in Thailand, *Sol. Energy Mat. Sol. Cells* **94** (2010), 1437–1440.

[112] P. J. Flory and J. Rehner, Jr, Statistical mechanics of cross-linked polymer networks ii. swelling, *J. Chemical Physics* **11**(11) (2004), 521–526.

[113] A.Standard, D2765-01 (reapproved 2006) standard test methods for determination of gel content and swell ratio of cross linked ethylene plastics, ASTM International, West Conshohocken, PA, 2006.

[114] Energiedispersive Röntgenspektroskopie, Wikimedia, 2013, available at: upload.wikimedia.org/wikipedia/commons/5/5a/Atom_model_for_EDX_DE.svg.

[115] Introduction to Energy Dispersive X-ray Spectrometry (EDS), University of California Riverside, Central Facility for Advanced Microscopy and Microanalysis, 2013.

[116] J. Haunschild, *Lumineszenz-Imaging – Vom Block zum Modul*, Fraunhofer Verlag, Stuttgart, 2012.

[117] ASTM, 1925, 1925–70 – Standard Test Method for Yellowness Index of Plastics, Pennsylvania, USA, 1925.

[118] C. Peike, *Degradation Analysis of the Encapsulation Material in Photovoltaic Modules by Raman Spectroscopy*, Stuttgart, Fraunhofer Verlag, 2015.

[119] ASTME313, Standard practise for Calculating Yellowness and Whiteness indices from Instrumentally Measured Color Coordinates, Pennsylvania, USA, 2000.

[120] D. C. Harris and M. D. Bertolucci, *Symmetry and Spectroscopy – An Introduction to Vibrational and Electronic Spectroscopy*, Oxford University Press, New York, 1978.

[121] N. S. Nielsen, D. N. Batchelder, and R. Pyrz, Estimation of crystallinity of isotactic polypropylene using Raman spectroscopy, *Polymer* **43** (2002), 2671–2676.

[122] Koenig, J.L., Infrared and Raman spectroscopy of polymers, *Rapra Review reports* **12** (1) (2001), 16–26.

[123] H. J. Bowley, D. L. Gerrard, and I.S. Biggin, The use of laser Raman spectroscopy to study the degradation of poly(vinyl chloride), *Polymer Degradation and Stability* **20**(34) (1988), 257–269.

[124] R. A. Mickiewicz, E. Cahill, and P. Wu, Non-destructive Determination of the Degree of Crosslinking of EVA Solar Module Encapsulation Using DMA Shear Measurements, Proceedings of IEEE, 2011.

[125] C. Peike, W. Phondongnok, T. Kaltenbach, K.-A. Weiss, and M. Koehl. Non-destructive determination of the cross-linking degree of EVA by Raman Spectroscopy, *Open Journal of Renewable Energy and Sustainable Development* **1** (1) (2014), 14–21.

[126] C. Peike, S. Hoffmann, I. Dürr, K. A. Weiss, and M. Köhl, The Influence of Laminate Design on Cell Degradation, *Energy Procedia* **38** (2013), 516–522.

[127] M. Koehl, Modelling of the nominal operating cell temperature based on outdoor weathering, *Solar Energy Materials and Solar Cells* **95**(7) (2011), 1638–1646.

[128] P. Huelsmann, M. Heck, and M. Koehl, M., Simulation of water vapor ingress into PV-modules under different climatic conditions, *J. Materials* (2013).

[129] F. C. Krebs, *The Different PV Technologies and How They Degrade, Stability and Degradation of Organic and Polymer Solar Cells*. JohnWiley & Sons, 2012.

[130] Fundamental Properties of Solar Cells and Pastes for Silicon Solar Cells. Stephenson and Associates, Inc., 2010.

[131] M. Koentges, V. Jung, and U. Eitner, Requirements on metallization schemes on solar cells with focus on photovoltaic modules, in: Proceedings of 2nd Workshop on Metallization for Crystalline Silicon Solar Cells, Constance, Germany, 2010.

[132] P. Huelsmann, K.-A. Weiss, and M. Koehl, Temperature-dependent water vapour and oxygen permeation through different polymeric materials used in PV-modules, Progress *in Photovoltaics: Res. Appl.* **22**(4) (2012), 415–421.

[133] E. E. Stansbury and R. Buchanan, Fundamentals of Electrochemical Corrosion. ASM Internat.,2000.

[134] T. Graedel and R. Frakental, *J. electrochem. Soc.* **137**(8) (1990), 2385.

[135] C. Heces, J. Alonso, and F. Corvo, *Rev. CENIC* **18**(2–3) (1987).

[136] K. Slamova, R. Glaser, C. Schill, S. Wiesmeier, and M. Koehl, Mapping atmospheric corrosion in coastal regions: Methods and results, *J. Photonics for Energy* **2**(1) (2012).

[137] G. Meira, Modelling sea-salt transport and deposition in marine atmosphere zone-a tool for corrosion studies, *Corrosion Science* **50**(9) (2008), 2724–2731.

[138] S. Oesch, M. Faller, Environmental effects on materials: The effect of the air pollutants SO_2, NO_2, NO and O_3 on the corrosion of copper, zinc and aluminum. A short literature survey and results of laboratory exposures, *Corrosion Science* **39**(9) (1997), 1505–1530.

[139] G. Jorgensen, personal correspondance, 2010.

[140] M. Lechner, K. Gehrke, and E. Nordmeier, *Makromolekulare Chemie*, Springer Verlag, Heidelberg, 2009.

[141] M. Köhl, D. Philipp, K.-A. Weiß, Rundvergleich von UV-Prüfeinrichtung für Photovoltaik-Module, GUS Jahrestagung 2011, Stutensee.

[142] M. Heck, M. Köhl, D. Philipp, J. Schlieper, and K.-A. Weiß, Investigation of surface temperatures, in: T. Reichert, T. (ed.), Natural and Artificial Ageing of Polymers: 3rd European Weathering Symposium. CEEES Publication No. 8, 2007, ISBN: 978-3-9810472-3-3.

[143] S. Hoffmann and M. Koehl, Effect of humidity and temperature on the potential-induced degradation, *Prog. in Photovolt: Res.Appl.* **22**(2) (2012), 173–179.

[144] J. Berghold, O. Frank, H. Hoehne, S. Pingel, B. Richardson, and M. Winkler, Potential induced degradation of solar cells and panels. 25th European Photovoltaic Solar Energy Conference and Exhibition / 5th World Conference on Photovoltaic Energy Conversion, 6–10 September 2010, Valencia, Spain.

[145] J. Bierbaum, D. Philipp, S. Stecklum, I. Dürr, and K.-A. Weiß, Investigation on Snail Track Formation, Degradation Mechanisms and Raman Spectroscopic Examination of the Corrosion Products, EU PVSec, 2015.

[146] Review of Failures of Photovoltaic Modules – Report IEA-PVPS T13-01:2014, Available at: http://www.isfh.de/institut_solarforschung/files/iea_t13_review_of_failures_of_pv_modules_final.pdf

[147] M. Köhl, B. Carlson, G. J. Jorgensen, and A. W. Czanderna, *Performance and Durability Assessment*, Elsevier, 2004.

Index

Symbols
2-diode model 23

A
absorptance 141
absorption 27
absorption coefficient 114
accelerated aging 183
accelerated aging test 193
acceleration factor 183
acceptor atoms 9
acoustic impedance 169
activator 51
active bypass diodes 82
active electronics 42
additives 74
adhesion strength 73
air mass 135
Al-BSF layer 126
alloy 54
AM 1.5 spectrum 15
ambient temperature (macroclimate) 173
angle-dependent reflectance 112
angular distribution 142
anisotropic conductive adhesives 61
annealed glass 65
annual yield 133
anti-reflective coating 38, 66, 113, 115, 128
anti-soiling coating 68
anti-Stokes–Raman scattering 163
architectural glass 65
arcing 82
area related cost 128
array of wires 52
Arrhenius equation 173
assembly line 84
Auger spectroscopy 157

B
back junction back contact 35
back surface field 34
back-contact 31
back-rail 83
back-tracking 132
backsheet 41, 69, 122

bandgap 10, 25
base 9
bathtub curve 198
beam irradiance 133
bifacial 6, 24, 31, 71, 94, 97
bifacial power gain factor 144
bifacial yield gain 145
binning 124
black body 26
black standard temperature 188
breakdown voltage 10, 80, 105
building integrated 7, 71
bulk absorptance 128
bulk transmittance 114
busbar 32, 125
bypass diode 41, 42

C
cables 126
CCD camera 99
cell interconnector 177
cell spacing 122
cell string 43
certification of PV module 197
chain scission 179, 185
change factor 107
characterisation routine 193
chemically strengthened 66
circuit parameters 11
clamps 83
clearness index 134
climatic chamber 186
coefficient of thermal expansion 30
cohesive solder fraction 59
commercial segment 7
conduction band 10
conductive adhesives 125
conductive filler 60
conductivity 11
contact angle 50
contact pads 31, 119
contact soldering 86
cooling phase 52
cooling press 89
cosine effect 139

creep 59
creep strength 59
cross-link 155
cross-linking reaction 179
CTM efficiency 128
CTM power 128
cyclic mechanical loads 59

D

daily diffuse horizontal 134
damp heat 199
damp-heat testing 193
dark IV curve 13, 97
declination angle 134
degradation process 173
delamination 169
depletion region 9
dewetting 51
dielectric breakdown 42, 69
dielectric rear side passivation 34
differential scanning calorimetry (DSC) 79, 155
diffuse component 133
diffuse sky irradiance 133, 134
diffusion barriers 42
diffusion current 9
diffusion voltage 9
diode 9
diode current 15
diode equation 15
distributed contacts 35
divided cell 126
double-junction 24
downholder 86
drift current 9
dynamic mechanical analysis 72, 156

E

edge contact 35
edge seal 76
EDX microanalysis 156
effective optical medium 66
effective width 117
efficiency 6, 14
elastic limit 44
elastomer 74
electric isolation 69
electrically conductive adhesives 60
electrochemical reaction 178
electroluminescence (EL) imaging 99, 160

electromagnetic radiation 172
element maps 63
elongation at fracture 44
emitter 9
encapsulant 41, 71
energy dispersive X-ray spectroscopy 63
energy gap 163
energy payback time 146
energy rating 96
epitaxial growth 30
equivalent circuit 11, 94
ethylene-vinyl acetate 74
eutectic 55
external load 171
external quantum efficiency (EQE) 25, 158
external spectral response 25

F

feed-in-tariff 5
fill factor 14
finger 31
flasher measurement 89, 158
float glass 64
fluorescence tube 185
flux 51, 85
forward bias 10
fracture energy 62
fracture pattern 62
fracture strain 44
frame 42, 82
frame tape 84
Fresnel equations 112, 140
front single side efficiency 94
front-to-back contact 31

G

gel content 74, 78, 155
geometrical width 117
glass 64
glass transition temperature 72
global horizontal 133
gradient index 66
grid-connected 6
ground-reflected irradiance 133

H

hail impact test 199
heat dissipation 82
hetero-junction 24

homologous temperature 59
honeycomb texture 38
hot-spot 13, 80, 105
hot-spot test 199
hour angle 133
humidity 174
humidity freeze 199
hydrolytic degradation 174
hydrophilic 69
hydrophobic 69

I

ideality factor 16
IEC 60904-9 186
IEC 61215 186, 195, 197
IEC 61646 198
inactive area 70, 110, 127
incidence angle 139
incidence angle modifiers 139
industrial segment 7
infant mortality 198
infrared soldering 86
InGaAs photodiode arrays 99
inhomogeneous irradiance 104
interconnector 31, 125
interdigitated back contact 35
intermetallic compound 55
internal quantum efficiency (IQE) 25, 158
inverted pyramid 12, 68
inverters 132
ionomer 74
irradiance 11, 139, 143
irradiation 132, 146
isothermal fatigue 59
isotropic conductive adhesives 60
isotropic texture 38
IV curve 11, 94
IV parameters 11, 94

J

joint 50
junction box 41, 80, 89

K

kerfless wafering 30
Kirchhoff's current law 16

L

Lambertian scattering 118
laminator 88
laminator pins 88
layup 84
lead 57, 92
lead-free 57
leakage current 42, 73
LED flasher 90
lift-off 30
light induced current 15
light induced degradation 24, 96
light weight 65
liquidus temperature 54
local solar time 134
low iron 65
low light 136
low light response 29
low solid 51
luminescence imaging 99, 159

M

maximum power point 14
mechanical impact 59
mechanical load 177
mechanical load testing 199
mechanical strength 30, 65
melting temperature 50, 55
metal corrosion 178
metal halide lamp 185
metal wrap through 35
metallic interdiffusion 52
metallization 31
metallography 62
micro-crack 30
micro-inverter 82
mismatch loss 124
mixed crystal 54
molecular bond 172
monofacial 12
multi-level laminator 89
multi-wire 46
MWT 48

N

n-type 9
nanoindentation test 167
Nernst equation 178
neutral plane 71

no-clean 51
nominal module operating temperature 138
nominal operating cell temperature 138

O
off-grid 6
open circuit voltage 13, 14
outdoor exposure 198
outdoor operation 131
oxygen 179

P
p-type 9
parallel resistance 15, 95
partial shading 42, 66, 80
patterned glass 65
peel angle 77
peel speed 77
peel test 62, 77, 158
percolation threshold 60
performance ratio 131
permeability 175
phase diagram 55
plasma texture 38
plastic deformation 59
plating 33
plugs 80
polymer degradation 179
polymer matrix 60
polymeric chain 173
polymeric degradation 174
potential induced degradation 73
power optimizer 82
power output 158
power temperature coefficient 136
preconditioning 96
preheating zones 86
pseudo-square 29, 110, 122
pulsed sun simulator 93
PV thermal 8

Q
quarter-wave layer 67

R
$R_{P0.2}$-value 44
radiant flux 14
radiation source 159
random pyramids 37
ray tracing 122
reactive ion etching 38
rear cover 69
rear side irradiance 144
rear single-side efficiency 94
recycling 91
reference spectrum 27
reflectance 68, 70, 112, 122
refractive index 37, 115, 117
residential segment 7
reverse bias 10, 104
reverse bias voltage 105
ribbon 44, 62
rolled glass 65
round wire 120
Rutherford–Bohr atom model 156

S
safety testing 200
sample temperature 188
sample temperature (microclimate) 173
saturation current 16
saw-tooth profile 45
scanning acoustic microscopy 169
scanning electric microscopy 63
scattering 27
Schottky diodes 81
series resistance 15, 43, 44, 61, 95, 125
service life 91
shading 132
shear stress 58, 72
shelf life 84
shingle-type string 50
short circuit current 13, 14
shunt resistance 15, 95
silicone 74
single side efficiency 94
single-junction 24
Snell's law 140
soiling 68, 132
solar altitude angle 133
solar azimuth angle 133
solar cell 9
solar constant 26, 134
solar coordinates 133
solar elevation angle 27
solar grade glass 65
solar simulator 93
solar transmittance 111

solder 50
solder paste 51
soldering 50
solid state diffusion 54
solidus temperature 54
solubility 54
Soxhlet apparatus 78
space charge region 9
spacial uniformity 90, 93
spectral irradiance 26
spectral mismatch factor 143
specularly 45
standard test conditions 14, 93
stationary mounted 131
steady state simulator 93
Stokes–Raman scattering 163
stress-strain curves 44
stringer 85
structured conductive layer 49
structured glass 141
structured reflector 70
substring 105
sunset hour angle 136
surface sensitivity 158
surface temperature 175
surface tension 51

T

tapered interconnector 47
temperature 173
temperature coefficient 28
temperature cycling 188, 199
temperature gradient 66
tensile strength 65
texture 117
texturing 37
thermal voltage 16
thermally strengthened glass 65
thermally toughened glass 65
thermo-mechanical fatigue 59
thermoplastic polyolefin elastomer 74
threshold voltage 10

time of wetness (TOW) 175
time to rupture 59
tin pad 34
total internal reflection 45, 117, 118
tracker mounted 131
transition temperature 55
transmission 167
transmittance 112, 140
trimming 89
typical meteorological year 133

U

utility segment 7
UV dose 173
UV preconditioning 198
UV testing 193
UV transmittance 115

V

valence band 10
vertical mounting 142
vibrational energy level 164
vibrations 59
void 59
volume resistivity 60, 72

W

wafer thickness 91
wafer-based 3
water vapor diffusion 175
water vapor transmission rate 69
weak organic acids 51
weight 71
wetting 50
white standard temperature 188

X

Xenon light source 185

Y

yield strength 44